TEXAS INSTRUMENTS

Activities for Algebra with the TI-73

Rachel E. Newman-Turner
Baltimore School for the Arts
Baltimore, MD

Robert S. Goodman
Hunter College High School
New York City, NY

Important notice regarding book materials

Texas Instruments Incorporated
7800 Banner Drive, M/S 3918
Dallas, TX 75251

Attention: Manager, Business Services

Printed in the United States of America.

ISBN: 1-886309-40-X

Chapter illustrations by Garrison Graphics.

We invite your comments and suggestions about this book. Call us at **1-800-TI-CARES** or send e-mail to **ti-cares@ti.com**. Also, you can call or send e-mail to request information about other current and future publications from Texas Instruments.

Visit the TI World Wide Web home page. The web address is: **education.ti.com**

Contents

Preface

This collection of activities is intended to provide middle and high school Algebra I students with a set of data collection investigations that integrate mathematics and science and promote mathematical understanding. The use of the TI-73 is an integral part of each activity. The activities are designed to be used in a mathematics classroom and require a minimum of equipment, setup, and materials. They are well suited to environments where collaborative teams of mathematics and science teachers plan interdisciplinary lessons. Each activity is based on a real-world problem situation and incorporates modeling of realistic data that encourages exploration of mathematical relationships and "what if" investigations. Many of the activities are intended for students working in groups of four or fewer.

These activities use technology to represent data graphically, explore patterns, and model the data. They also use technology to assist students in analyzing and interpreting the data in order to make decisions. The activities are keystroke-intensive and can be used by teachers and students whose technology skills range from beginner to advanced ability. They are designed to stand alone, and can be done in any order. The length of time for the completion of each activity ranges from approximately 45 minutes to 120 minutes.

Each activity consists of:

- **Objectives:** A brief statement of the activity objectives.
- **Materials:** A list of the materials needed for the activity.
- **Introduction:** A brief description of the situation that serves to motivate the student to investigate the problem.
- **Problem:** A statement of the problem to be solved.
- **Data Collection Instructions:** The specific steps a student or a team of students should follow to complete the activity. This section contains detailed instructions with keystrokes and sample screens. One copy of the instructions per group is sufficient.
- **Data Collection and Analysis pages:** This section is for data collection and analysis and includes questions for the students. Each student should be given a copy.
- **Extensions:** Investigations that provide opportunities for more advanced study on the topic covered in the activity. Some activities do not contain this section.
- **Teacher Notes:** Designed for teacher use. This section provides additional teaching and calculator tips.
- **Answers to Analysis Questions:** Answers and sample data for the Data Collection and Analysis pages.

It is our sincere hope that the graphing activities in this book will provide a catalyst for stimulating mathematical explorations and investigations. We hope that you and your students enjoy working on the activities as much as we enjoyed writing them.

We would like to thank the following people for their time, ideas, and support. We greatly appreciate the contributions of:

- Tobie Brandriss, Yetzchaq Eaton, Madeline Forrest, David Hankin, Dr. Martin Kornblatt, Cheryl Newman-Pope, Don Porter, Aaron Pross, Asumana Randolph, Cecil Bernard Tucker, Sr., and our colleagues at Texas Instruments who assisted in the preparation of this book.
- The students at the Baltimore School for the Arts and Hunter College High School for their enthusiasm for learning and exploring technology.

— *Rachel E. Newman-Turner*

— *Robert S. Goodman*

About the Authors

RACHEL E. NEWMAN-TURNER is a mathematics teacher at the Baltimore School for the Arts in Baltimore, Maryland. She was the recipient of The Presidential Award for Excellence in Mathematics Teaching in 1996, the State of Maryland Governor's Citation for Excellence as an Exemplary Educator in Maryland, 1996, The Tandy Technology Scholar – Outstanding Teacher for Academic Excellence in Mathematics in 1997, Baltimore City Teacher of the Year in 1997, Maryland Teacher of the Year Finalist in 1998, and was a national Finalist for The Walt Disney Company Presents the American Teacher Award for Mathematics in 1998. Rachel has authored Texas Instruments first online course entitled *Algebra Using the TI-83 Plus.* She has co-authored an Exploration book entitled *Activities for Algebra with the TI-83 Plus.* She has written materials for the successful series *Life By the Numbers* produced by PBS and Texas Instruments, authored a publication entitled *Mathtrails* produced by PBS and Texas Instruments, and taped video lessons for *PBS MATHLINE High School Project – Focus on Algebra.* She was a national question writer for *MATHCOUNTS.* She served on the Content Validation Committee for the National Board for Professional Teaching Standards (NBPTS); and testified before the United States Senate caucus on *Science and Technology Education: The Weak Link in Building a World Class Workforce.* Rachel is a Teachers Teaching with Technology (T^3) instructor and has written summer institute materials. She has conducted technology workshops on the use of graphing calculators and the Calculator-Based Laboratory (CBL) system in various school districts across the nation for teachers, principals, and school administrators. She presents workshops and demonstrations on using technology to enhance the teaching of mathematics at national and state conferences such as the National Council of Teachers of Mathematics (NCTM) and Teachers Teaching with Technology (T^3).

ROBERT S. GOODMAN is the chairman of the science department at the North Shore Hebrew Academy High School in Great Neck, New York. He was the recipient of a Sci-Mat Fellowship from the Council on Basic Education/NSTA in 1993, the Entomological Society of America Award for Secondary Education in 1994, was chosen as a Genentech Access Excellence Fellow in 1995 and received a RadioShack National Teacher Award in 2001. He has written a book, *Calculator Based Biology-A Biology Laboratory Manual Using Probeware and Graphing Calculators* (1996), an article for American Biology Teacher (*The Animal Communication Treasure Hunt* - 1996), another Explorations Book, *Activities for Algebra with the TI-83 Plus,* and has co-hosted a World Wide Web session on using computers, CBLs, and probeware for Genentech's Access Excellence. Robert is a Teachers Teaching with Technology (T^3) instructor and has written summer institute materials. He presents workshops and demonstrations at national and regional conferences including National Association of Biology Teachers (NABT), National Science Teachers Association (NSTA), and Teachers Teaching with Technology (T^3).

Contents and Calculator Functionality Checklist

EXPLORATIONS

Activity 1

Cricket Thermometers

Objectives

- To investigate the relationship between temperature and the number of cricket chirps
- To find the *x* value of a function, given the *y* value
- To find the *y* value of a function, given the *x* value
- To use technology to find a linear regression
- To use technology to plot a set of ordered pairs

Materials

- TI-73 graphing device
- Cassette player
- Tape of crickets chirping

Introduction

Anyone who has ever been out in the country on a summer evening is familiar with the chirp of a cricket. Chirping patterns convey different messages and are different from species to species. Male crickets chirp in order to attract and court female crickets, and to stake a claim to their territory. The mature male cricket makes the sound by rubbing together his forewings, much like a violinist produces sound by rubbing a bow along the strings of his instrument.

Crickets are insects, and like other insects, they are ectothermic, which means that their body temperature rises or falls when the temperature of the environment rises or falls. The metabolism of an insect fluctuates with its body temperature.

Is a cricket's rate of metabolism reflected in the frequency of its chirps? In this activity, you will graphically analyze the relationship between cricket chirps and temperature.

Problem

How can you use the frequency of cricket chirping to predict the temperature of a habitat?

Collecting the data

1. Your teacher will play a tape of cricket chirps at different temperatures. The narrator will give the temperature and then you will listen to the cricket chirping. Your teacher will tell you when to start counting and when to stop.
2. Record the temperature and the number of chirps in 15-second intervals on the **Data Collection and Analysis** page.

3. Your teacher will play a tape of three crickets, each chirping at *unknown* temperatures. Record the number of chirps in 15-second intervals and then multiply by 4 in order to determine the number of chirps per minute. Record the number of chirps on the **Data Collection and Analysis** page.

Setting up the TI-73

Before starting your data collection, make sure that the TI-73 has the STAT PLOTS turned OFF, Y= functions turned OFF or cleared, the MODE and FORMAT set to their defaults, and the lists cleared. See the Appendix for a detailed description of the general setup steps.

Entering the data in the TI-73

1. Press [LIST].

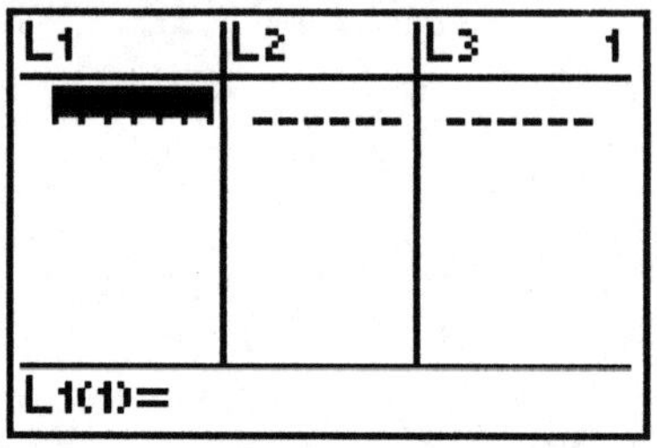

2. Enter the temperatures in **L1**.
3. Enter the number of chirps per 15-second intervals in **L2**. (Make sure your pairs of temperature and chirps match in each column.)

   ***Note**: Your cricket chirp count might differ from what is shown.*

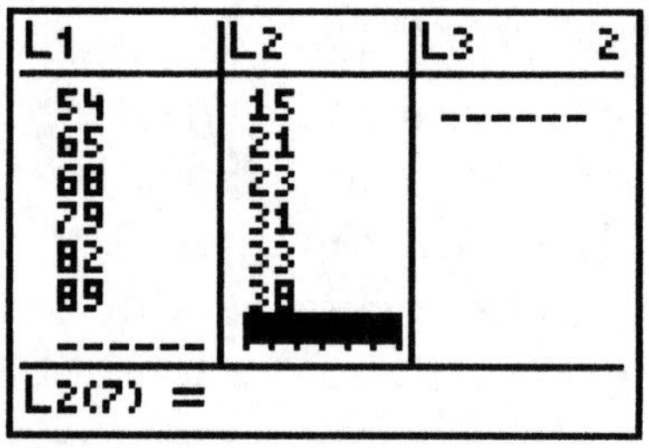

4. Press [2nd] [STAT] [▸] to move the cursor to the **OPS** menu.

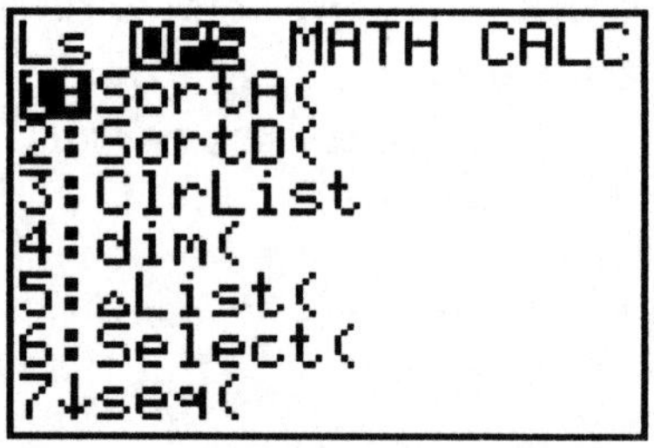

5. Select **1:SortA(** by pressing **1** or [ENTER].

6. Press [2nd] [STAT] **1:L1** [,] [2nd] [STAT] **2:L2** [)].

7. Press [ENTER]. The lists are sorted.

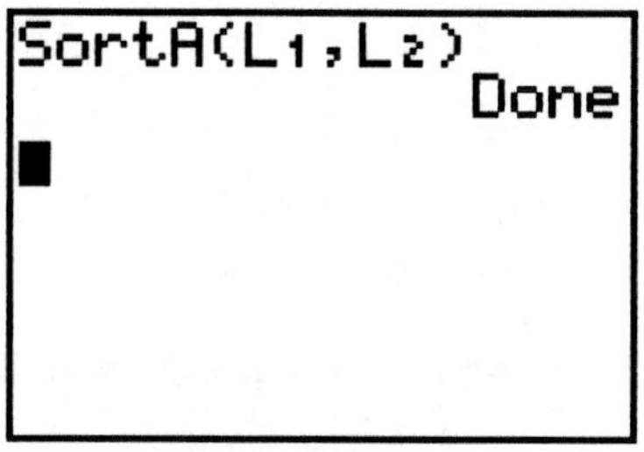

8. Press [LIST] to return to the data lists.

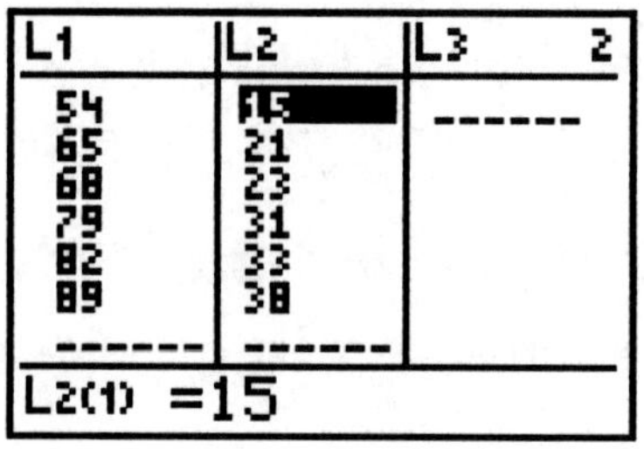

L1
L2
L3
2
54 65 68 79 82 89
15 21 23 31 33 38
L2(1) =15

You counted the number of chirps in 15-second intervals, but you would like to plot the number of chirps per minute. In order to get the number of chirps per minute, you must multiply all the entries in **L2** by 4.

9. Press [▶] and [▲] to highlight **L3**.

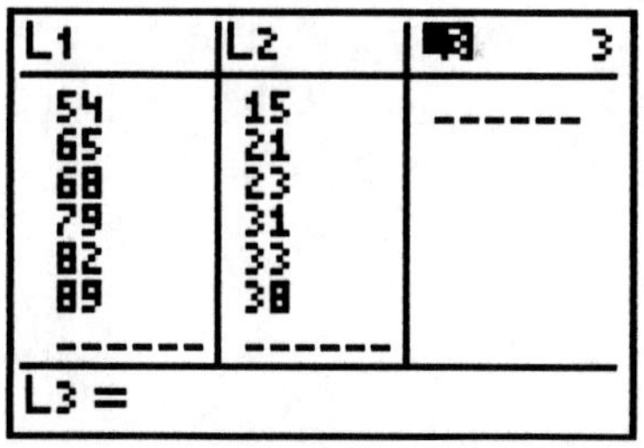

L1
L2
3
54 65 68 79 82 89
15 21 23 31 33 38
L3 =

10. Press [2nd] [STAT] **2:L2** [×] **4**.

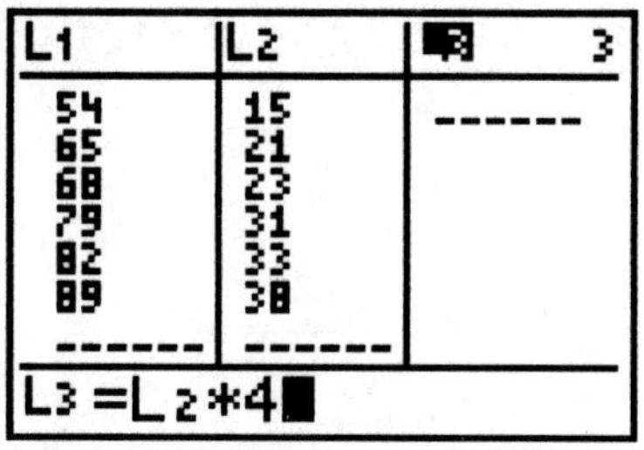

L1
L2
3
54 65 68 79 82 89
15 21 23 31 33 38
L3 =L2*4

11. Press [ENTER] to see the calculation.

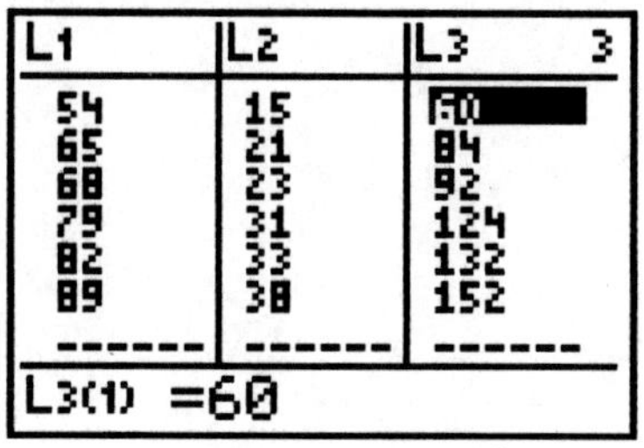

L1
L2
L3
3
54 65 68 79 82 89
15 21 23 31 33 38
84 92 124 132 152
L3(1) =60

Setting up the window

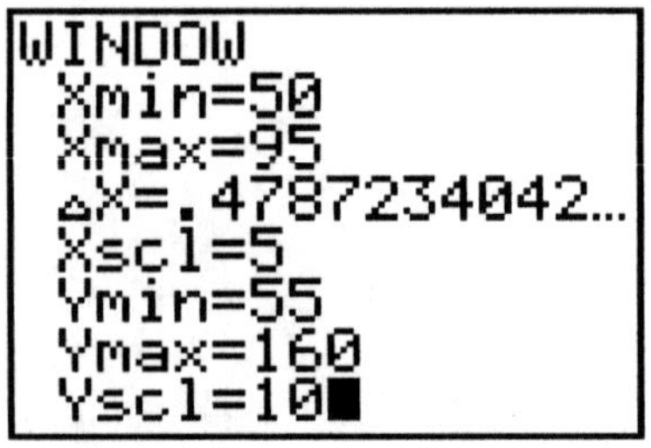

1. Press [WINDOW] to set up the proper scale for the axes.

2. Set the **Xmin** value by identifying the minimum value in **L1**. Choose a number that is less than the minimum.

3. Set the **Xmax** value by identifying the maximum value in each list. Choose a number that is greater than the maximum. **Do Not Change the ΔX Value.** Set the **Xscl** to **5**.

4. Set the **Ymin** value by identifying the minimum value in **L3**. Choose a number that is less than the minimum.

5. Set the **Ymax** value by identifying the maximum value in **L3**. Choose a number that is greater than the maximum. Set the **Yscl** to **10**.

Graphing the data: Setting up a scatter plot

1. Press [2nd] [PLOT]. Select **1:Plot1** by pressing **1** or [ENTER].

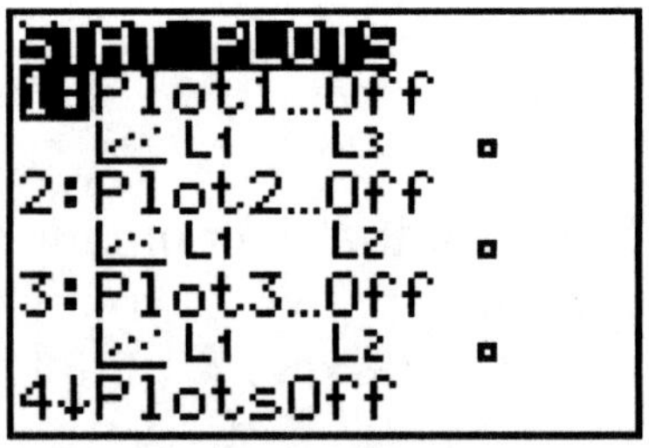

2. Set up the plot as shown by pressing [ENTER] [▼] [ENTER] [▼] [2nd] [STAT] **1:L1** [▼] [2nd] [STAT] **3:L3** [▼] [ENTER].

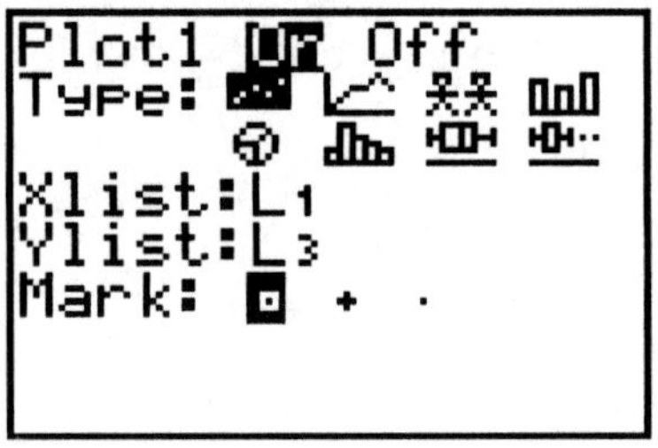

3. Press [GRAPH] to see the plot.

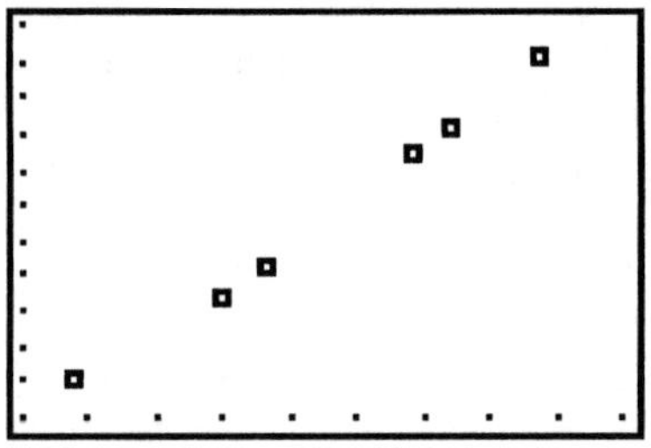

Analyzing the data

Finding a linear regression

1. Press [2nd] [STAT]. Press [◄] to move the cursor to the **CALC** menu.

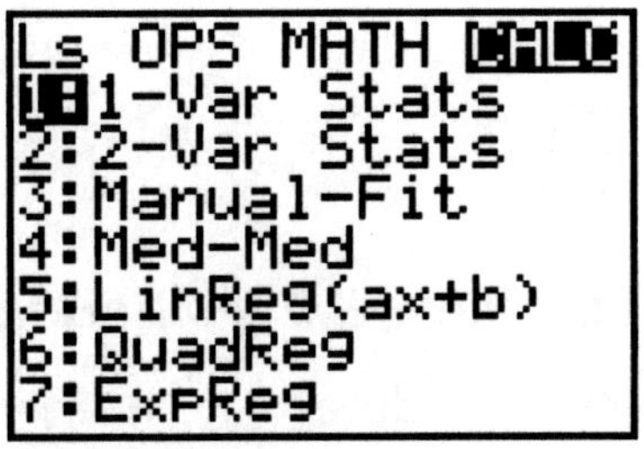

2. Select **5:LinReg(ax+b)** by pressing **5**.

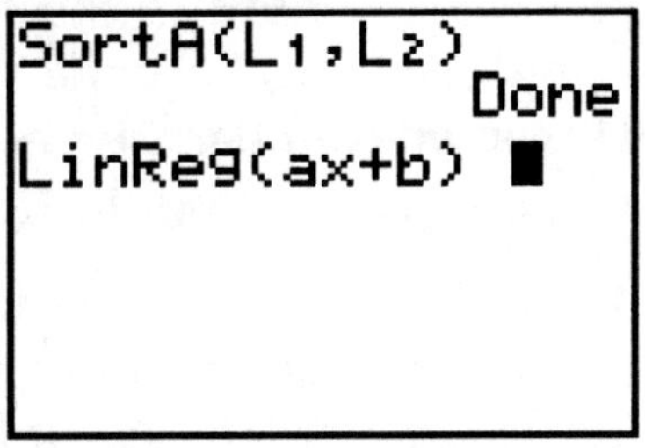

3. Press [2nd] [STAT] **1:L1** [,] [2nd] [STAT] **3:L3** [,].

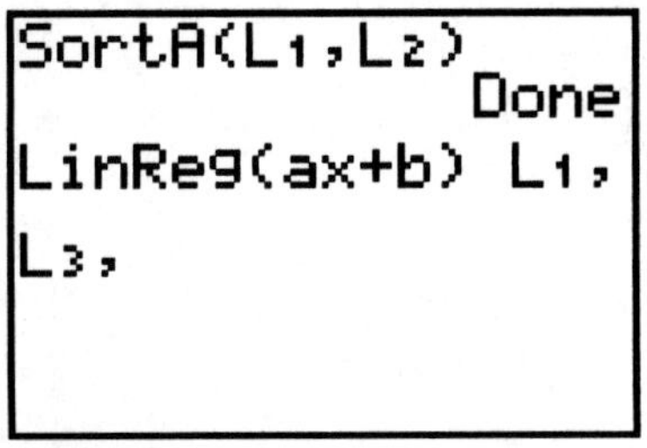

4. Press [2nd] [VARS]. Select **2:Y-Vars** by pressing **2**.

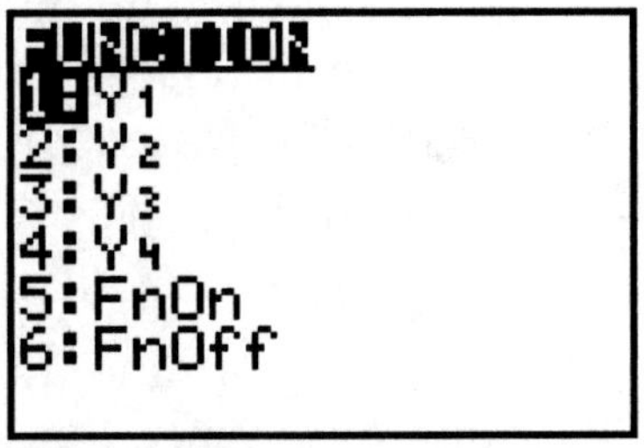

5. Select **1:Y1** by pressing **1** or [ENTER].

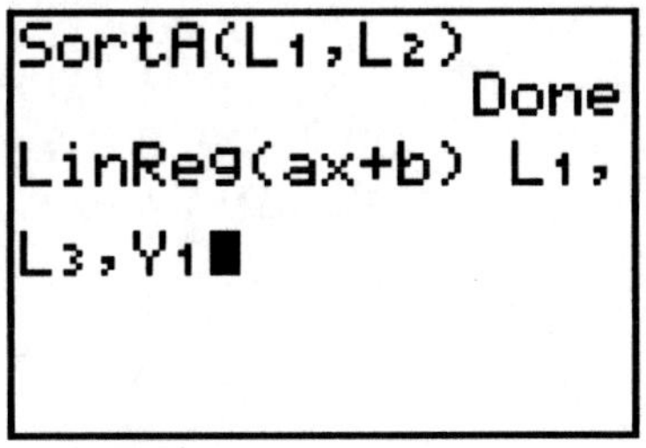

6. Press [ENTER] to calculate the linear regression. The function is pasted in **Y1**.

   ***Note**: Your values might differ from what is shown.*

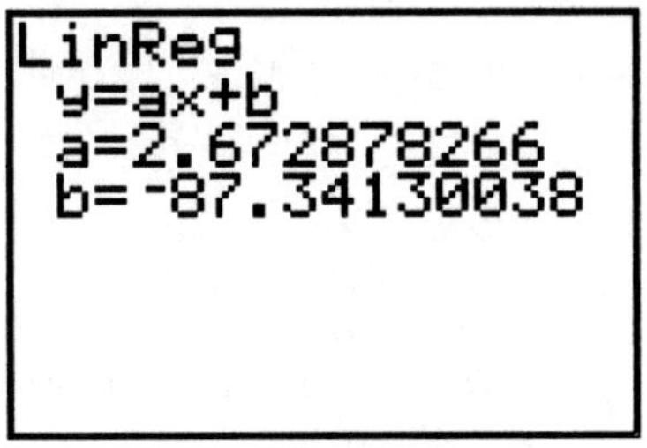

7. Press [GRAPH] to see the linear regression model.

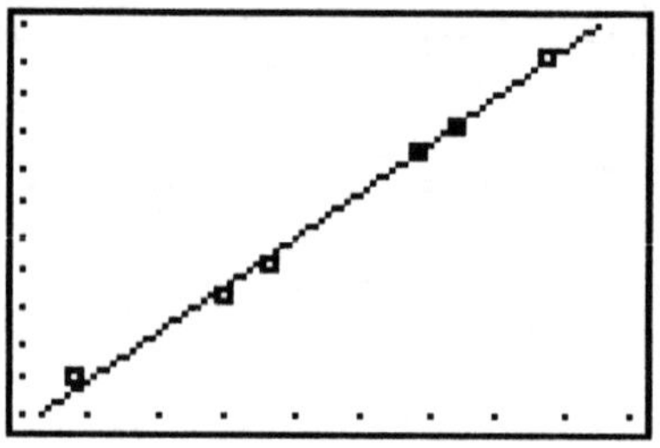

Determining the temperature of a habitat

You can determine the temperature of a habitat based on the number of cricket chirps per minute. In the example shown below, the cricket chirped 124 times per minute. You will use the actual number of cricket chirps per minute for Unknown Number 2 based on the tape recording that **you** listened to, or data given to you by your instructor.

1. Press [Y=]. Press [▼] until you are at **Y2**. Enter the number of chirps per minute for your unknown (124 in our example).

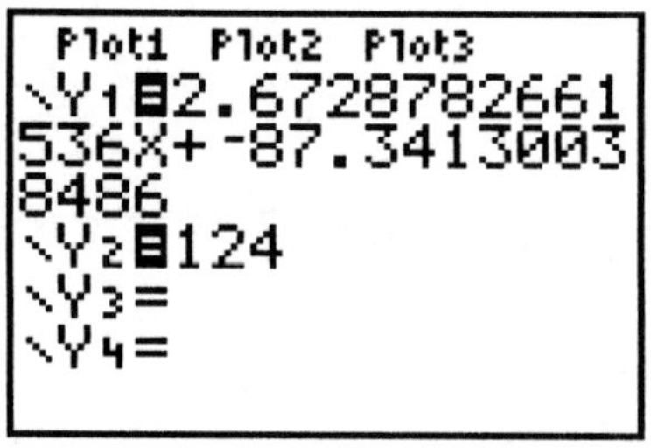

2. Press [GRAPH] to see the intersection of the two lines.

   The *x* value of the point where the two functions intersect is the temperature of the habitat where your *unknown* cricket was chirping.

   The table function of the TI-73 will be used to determine the coordinates of the point of intersection.

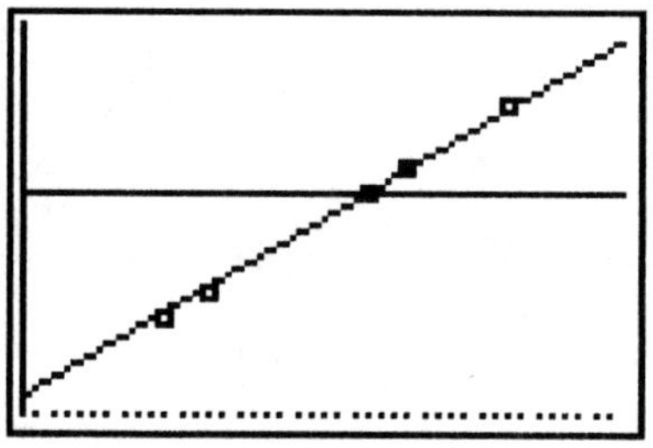

3. Press [2nd] [TBLSET]. Type in the lowest *x* value in **L1** (55 in the example). Press [▼] **5** to set the **ΔTbl** value.

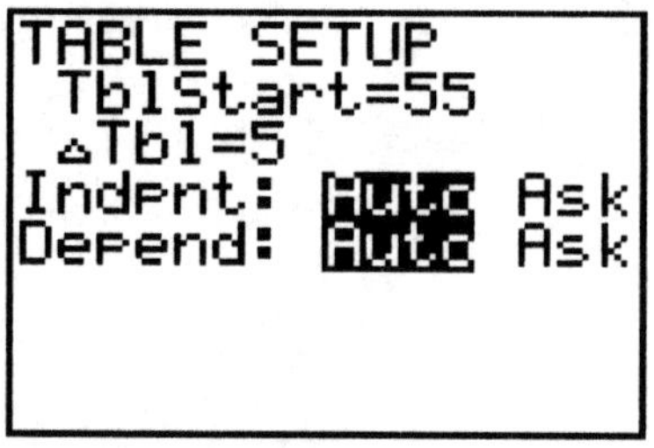

4. Press [2nd] [TABLE]. If necessary, use [▼] and [▲] to scroll the table.

   Note: *For this example, in the Y1 column, 124 chirps per minute falls between 113.12 and 126.49, which corresponds to 75 and 80 degrees in the X column. Based on that information, the table will be readjusted.*

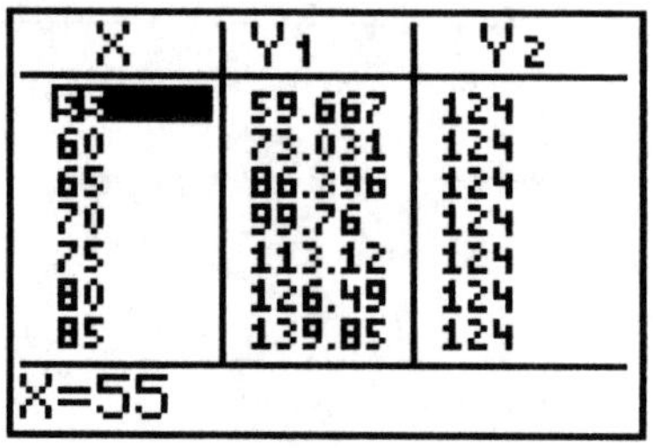

X Y1 Y2
55 59.667 124
60 73.031 124
65 86.396 124
70 99.76 124
75 113.12 124
80 126.49 124
85 139.85 124
X=55

5. Press [2nd] [TBLSET]. Enter your results from Step 4 for **TblStart**. Press [▼] **1** to set the **ΔTbl** value.

6. Press [2nd] [TABLE]. If necessary, use [▼] and [▲] to scroll the table.

   Note: *For this example, in the Y1 column, 124 chirps per minute falls between 123.82 and 126.49, which corresponds to 79 and 80 degrees in the X column. Based on that information, the table will be readjusted again.*

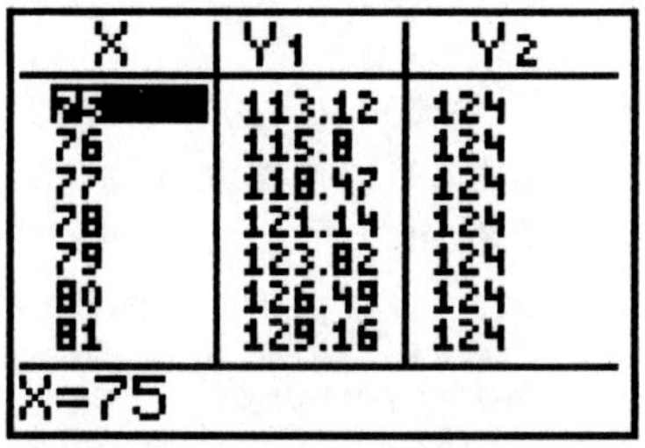

7. Press [2nd] [TBLSET]. Enter your results from Step 6 for **TblStart**. Press [▼] **0.1** to set the **ΔTbl** value.

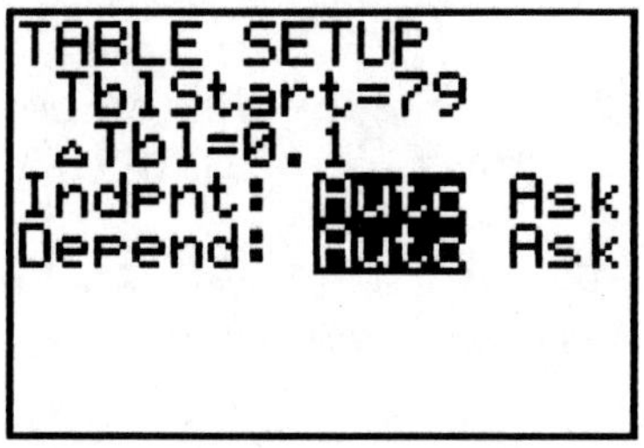

8. Press [2nd] [TABLE]. If necessary, use [▼] and [▲] to scroll the table.

   Note: *The data used to construct the linear model had temperatures measured to the nearest degree. Therefore, the unknown is determined to the same level of precision. From the table, 124 chirps per minute falls between 123.82 and 124.08, which corresponds to 79 and 79.1 degrees. Rounding to the nearest degree, the intersection point will be (79, 124).*

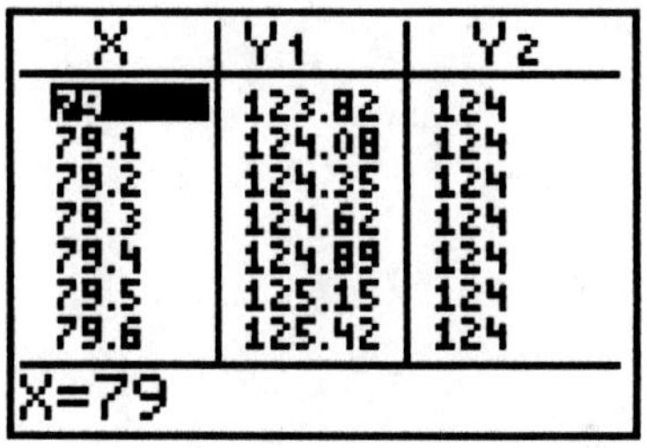

   To verify the coordinates graphically, we will use the **DRAW** function. Press [DRAW].

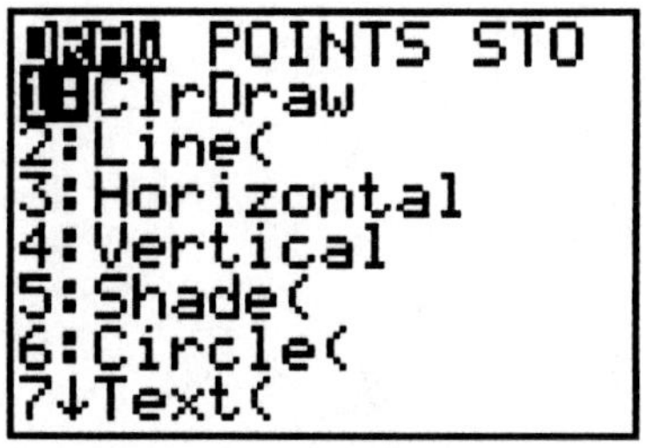

9. Select **4:Vertical** by pressing **4**.

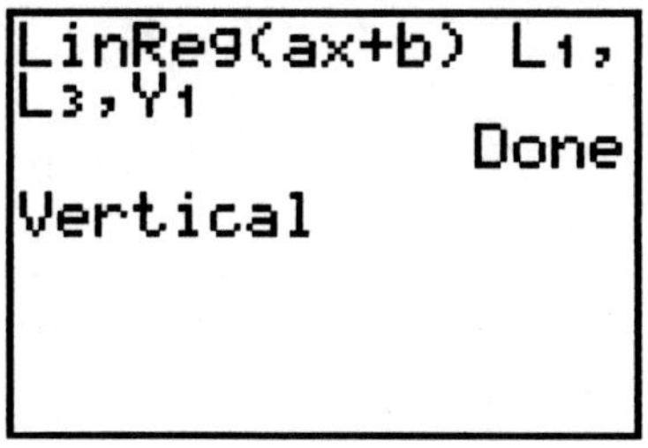

10. Type the results from Step 8. (In this example, 79.) Press [ENTER].

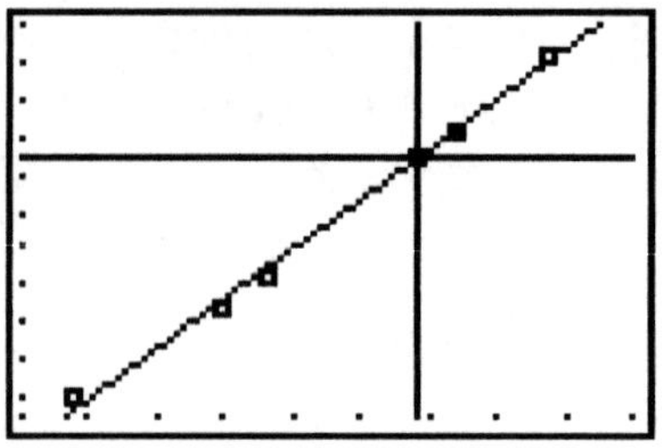

For this example, note that the coordinates of the point on the linear model where all of the lines intersect is defined by the vertical drawn at $x=79$ and the horizontal at $y=124$.

11. The coordinates of the intersection can be added onto your screen by pressing [DRAW] **7:Text**, moving the cursor near the point of intersect, and typing the results.

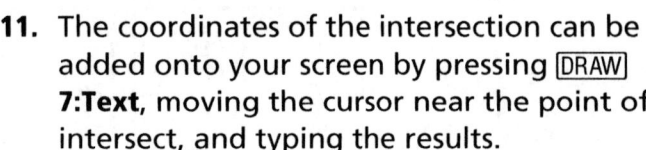

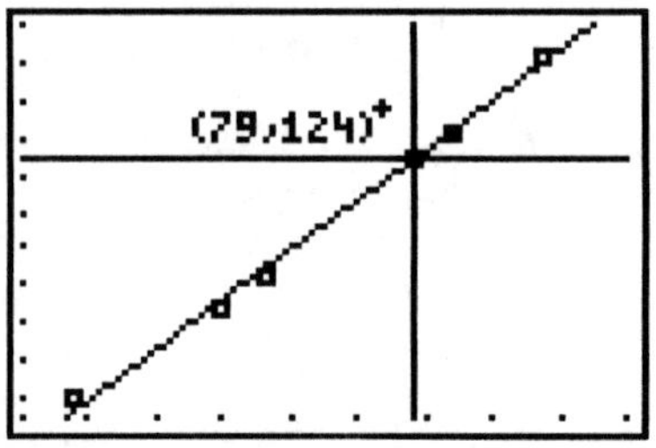

Note: *Text appears below and to the right of the cursor.*

12. Determine the temperatures of the habitats for all three unknown crickets.

Answer the questions on the **Data Collection and Analysis** page.

Data Collection and Analysis

Name ______________________

Date ______________________

Activity 1: Cricket Thermometers

Collecting the data

Temperature (°F)	Number of chirps per 15 seconds

	Temperature (°F)	Number of chirps per 15 seconds	Number of chirps per minute
Unknown Cricket Number 1			
Unknown Cricket Number 2			
Unknown Cricket Number 3			

Analyzing the data

1. Coordinates for the intersection of the two functions for Unknown Cricket Number 2:

   x = __________ *y* = __________

2. The *slope* of the linear regression line is ______________________

3. Explain what the *slope* represents in context with the data that you analyzed.

   ____

   ____

   ____

4. What does the *y* value of the intersection of the two functions represent?

   ____

   ____

5. What does the x value of the intersection of the two functions represent?

____

____

6. Determine the temperature of the habitat for unknown crickets 1 and 3. Repeat the procedure in the **Determining the temperature of a habitat** section.

Note: *For Unknown Cricket Number 1, you will have to adjust the window.*

Temperature of the habitat for Unknown Cricket Number 1: ____________

Temperature of the habitat for Unknown Cricket Number 3: ____________

7. You had to interpolate to determine the temperature of the habitat for Unknown Crickets number 2 and 3. *Interpolation* means to make a prediction *within* the bounds of known data. The key word is *within*. You had to extrapolate to determine the temperature of the habitat for Unknown Cricket number 1. *Extrapolation* means to make a prediction that a trend will continue *outside* the bounds of known data. From a scientific standpoint which is riskier, interpolation or extrapolation? Explain.

____

____

____

____

Extensions

- Predict the number of cricket chirps if the temperature is known. Suppose that the temperature of a habitat is 70°F. Explore how to predict the number of cricket chirps per minute.

  How many chirps per minute did you predict when the temperature of the habitat is 70°F?

- Based on the data, how many chirps per minute would a cricket make if the temperature of the habitat were 32°F or 212°F? (Remember to adjust the window.)

- Based on common sense, would you give the same answer? Explain.

Teacher Notes

Activity 1

Cricket Thermometers

Objectives

- To investigate the relationship between temperature and the number of cricket chirps
- To find the *x* value of a function, given the *y* value
- To find the *y* value of a function, given the *x* value
- To use technology to find a linear regression
- To use technology to plot a set of ordered pairs

Materials

- TI-73 graphing device
- Cassette player
- Tape of crickets chirping
- Stopwatch (optional)

Preparation

- You can purchase the audiocassette tape for this activity by writing to:

  Robert Anderson, Ph.D.
  Department of Biological Sciences
  Idaho State University
  Pocatello, ID 83209

  Ask Dr. Anderson to send and bill you for *Myths and Science of Cricket Chirps*. The cassette comes with a booklet.

Management

- When playing the tape, play the beginning of Side B. There is approximately 25 – 30 seconds of chirping at each temperature. Play the tape for a few seconds. Tell the students to start timing the chirps. Mark time for 15 seconds and then tell the students to stop. You may want to use a stopwatch. Ask the students to record the data on the **Data Collection and Analysis** page. The students will learn to use the spreadsheet functions of the TI-73 to determine the number of chirps per minute.

  ***Note**: As of the time of printing, there is a slight error on the tape. The narrator does not say when unknown cricket number 2 starts, although there is suddenly a clear change in the frequency of cricket chirps.*

- You may want to have the students transfer this data to one of the computer graphing programs (such as TI InterActive!™ software) and then produce a hard copy of their graphed data.

Answers to Data Collection and Analysis questions

Collecting the data

Sample sorted data:

Temperature (°F)	Number of chirps per 15 seconds
54	15
65	21
68	23
79	31
82	33
89	38

	x-coordinate (temperature — °F)	*y*-coordinate (chirps per minute)
Unknown Cricket Number 1	49	44
Unknown Cricket Number 2	79	124
Unknown Cricket Number 3	73	108

Analyzing the data

1. *Answers will vary.*

2. The *slope* of the linear regression line is __________.

   The slope of the linear regression line is 2.67.

3. Explain what the *slope* represents in context with the data that you analyzed.

   Slope is defined as rise over run. It represents how the number of chirps per minute changes as the temperature changes. In this problem, the slope is $\frac{2.67}{1}$ *or 2.67. This means that for every increase in temperature of one degree Fahrenheit, the number of cricket chirps per minute increases about 2.67.*

4. What does the *y* value of the intersection of the two functions represent?

   The y value represents the number of chirps per minute for the unknown cricket.

5. What does the *x* value of the intersection of the two functions represent?

   The x value represents the temperature that corresponds to the number of chirps per minute for the unknown cricket.

6. Determine the temperature of the habitat for unknown crickets 1 and 3. Repeat the procedure in the **Determining the temperature of a habitat** section.

   Note: *For Unknown Cricket Number 1 you will have to adjust the window.*

   For Unknown Cricket Number 1, the temperature is 49 degrees Fahrenheit.

   For Unknown Cricket Number 3, the temperature is 73 degrees Fahrenheit.

7. You had to interpolate to determine the temperature of the habitat for unknown crickets number 2 and 3. *Interpolation* means to make a prediction *within* the bounds of known data. The key word is *within*. You had to extrapolate to determine the temperature of the habitat for unknown cricket number 1. *Extrapolation* means to make a prediction that a trend will continue *outside* the bounds of known data. From a scientific standpoint, which is riskier, interpolation or extrapolation? Explain.

   *Extrapolation is riskier as it assumes a trend will continue beyond the experimental data. See answers to **Extensions** section.*

Answers to Extensions questions

- How many chirps per minute did you predict when the temperature of the habitat is 70°F?

  If the temperature is 70°F, then the cricket will chirp approximately 100 times per minute.

- Based on the data, how many chirps per minute would a cricket make if the temperature of the habitat were 32°F or 212°F? (Remember to adjust the window.)

  According to the graph, if the temperature were 32°F the cricket would chirp -1.8 times. If the temperature were 212°F, then the cricket would chirp 479 times. However, you know that frozen and boiled crickets do NOT chirp. The lesson here is that the graph is useful when you interpolate, for it is linear within the bounds of the known data. It is therefore reasonable to interpolate.

- Based on common sense, would you give the same answer? Explain.

  As common sense would dictate, extrapolation is another story. As the temperature is significantly increased or decreased, you move beyond the range of temperatures that the cricket could tolerate. The graph loses its linearity.

EXPLORATIONS

Activity 2

Follow the Golden Rule

Objectives

- To use technology to find ratios
- To use technology to find measures of central tendency
- To use technology to plot a box-and-whisker plot

Materials

- TI-73 graphing device
- Metric tape measure (meter stick)

Introduction

What could the *Mona Lisa* painting, sunflowers, pine cones, the family tree of the drone bee, the Great Pyramid of Giza, and the human body have in common? The answer is the *Golden Ratio*. Early Greek mathematicians were fascinated by this ratio. Euclid, the Greek mathematician, showed how to *divide a line in mean and extreme ratio*, which is called *finding the golden section G point on the line.* This means that *the ratio of the smaller part of a line (GB) to the larger part (AG) is equal to the ratio of the larger part (AG) to the whole line (AB)*. This ratio is approximately 1.618033989. The exact value of the Golden Ratio is $\frac{1+\sqrt{5}}{2}$.

$$\frac{GB}{AG} = \frac{AG}{AB} \quad \text{or} \quad \frac{1-x}{x} = \frac{x}{1}$$

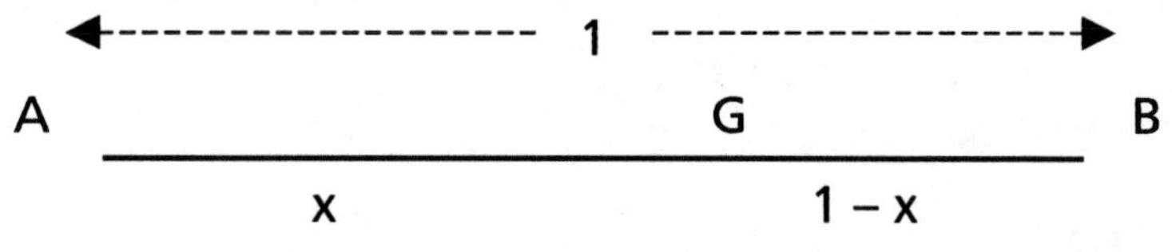

The Golden Ratio is said to be one of the most visually pleasing geometric forms. Masterpieces from ancient times as well as more recent works of art include examples of the Golden Ratio. A golden spiral and the *Fibonacci sequence* are closely related to the Golden Ratio and can be found in sunflowers and pine cones. The family tree of the drone bee can be linked to the *Fibonacci sequence*, which can be used to find the Golden Ratio.

The *Rhind Papyrus,* dating from 1650 B.C., is one of the oldest mathematical records in existence, giving evidence that the Egyptians had knowledge of the Golden Ratio or, as they referred to it, the *Sacred Ratio*. The Egyptians used the Golden Ratio when building the pyramids, temples, and tombs. Egyptian history shows how proportions of the human figure are related to the width of the palm of the hand. These measurements are based on the *Golden Ratio*. For example, the Egyptians believed the height of a person from the feet to the hairline was equal to eighteen palms. Is your height equal to eighteen of your palms?

Problem

Were the Egyptians correct in relating the Golden Ratio to the human body? Are the proportions in your body related to the Golden Ratio?

Collecting the data

1. You should have at least one partner for this activity. Obtain a tape measure from your teacher. Tape the tape measure to the wall. Make sure the tape measure has the lowest measurement starting from the floor.

2. Measure (a) the height of your partner; (b) the distance from your partner's navel to the floor; and (c) the distance from the top of your partner's head to his/her navel. Record these values in the table on the **Data Collection and Analysis** page.

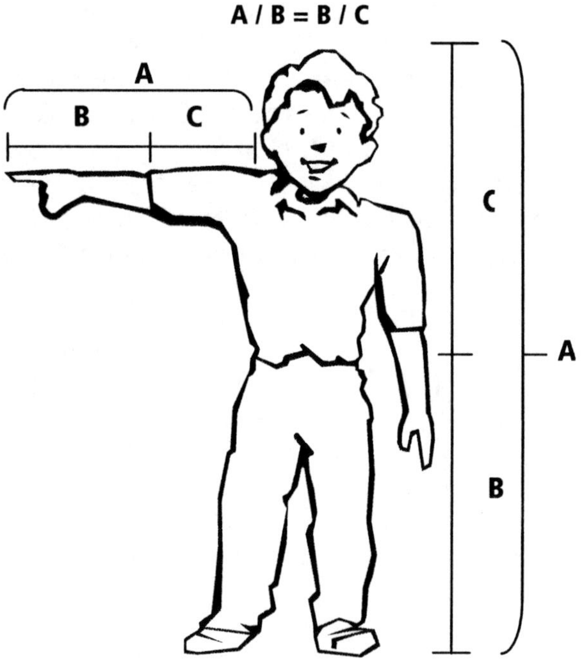

3. Measure (a) the distance from your partner's shoulder to the tip of his/her hand; (b) the distance from your partner's elbow to the tip of his/her hand; and (c) the distance from your partner's shoulder to his/her elbow. Record these values in the table on the **Data Collection and Analysis** page.

Setting up the TI-73

Before starting your data collection, make sure that the TI-73 has the STAT PLOTS turned OFF, Y= functions turned OFF or cleared, the MODE and FORMAT set to their defaults, and the lists cleared. See the Appendix for a detailed description of the general setup steps.

Entering the data in the TI-73

1. Press [LIST].

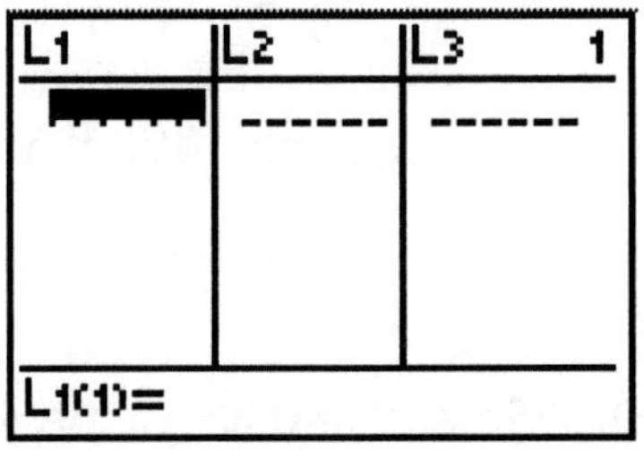

2. Enter the data from number 2 (a), (b), and (c) of the **Collecting the data** section in **L1**, **L2**, and **L3** respectively.

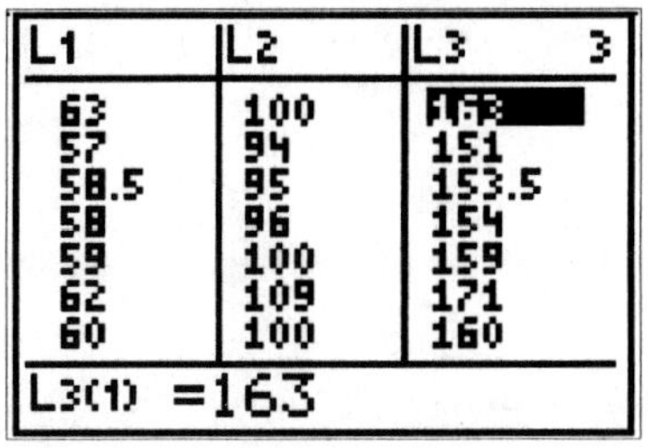

L1 L2 L3 3
63 57 58.5 58 59 62 60
100 94 95 96 100 109 100
151 153.5 154 159 171 160
L3(1) =163

3. To find the ratio of **L2** to **L1**, press [▶] [▲] to highlight **L4**. Press [2nd] [STAT] **2:L2** [÷] [2nd] [STAT] **1:L1**.

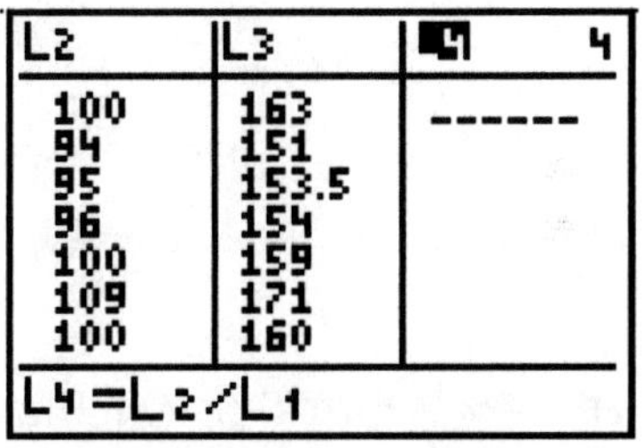

L2 L3 4
100 94 95 96 100 109 100
163 151 153.5 154 159 171 160

L4 =L2/L1

4. Press [ENTER] to complete the ratio calculation. The list is displayed as shown.

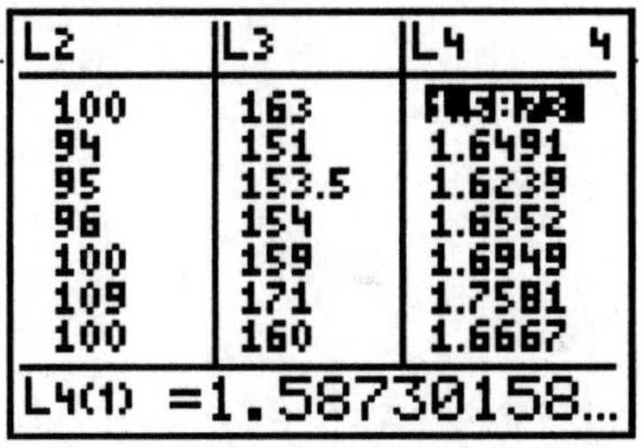

L2 L3 L4 4
100 94 95 96 100 109 100
163 151 153.5 154 159 171 160
1.6491 1.6239 1.6552 1.6949 1.7581 1.6667
L4(1) =1.58730158...

5. To find the ratio of **L3** to **L2**, press [▶] [▲] to highlight **L5**. Press [2nd] [STAT] **3:L3** [÷] [2nd] [STAT] **2:L2**.

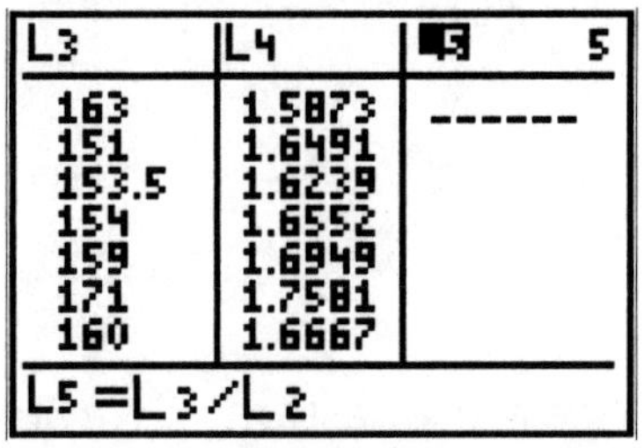

L3 L4 5
163 151 153.5 154 159 171 160
1.5873 1.6491 1.6239 1.6552 1.6949 1.7581 1.6667

L5 =L3/L2

6. Press [ENTER] to complete the ratio calculation. The list is displayed as shown.

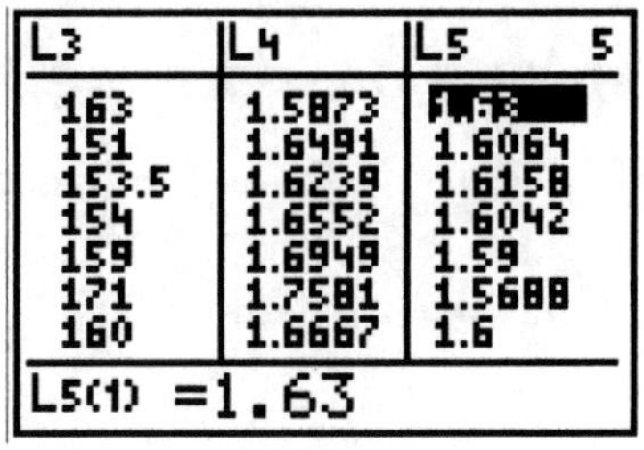

L3 L4 L5 5
163 151 153.5 154 159 171 160
1.5873 1.6491 1.6239 1.6552 1.6949 1.7581 1.6667
1.6064 1.6158 1.6042 1.59 1.5688 1.6
L5(1) =1.63

7. Find the mean of the class data for **L4**. Press [2nd] [QUIT] to return to the Home screen. Press [2nd] [STAT] [▶] [▶] to move the cursor to the **MATH** menu.

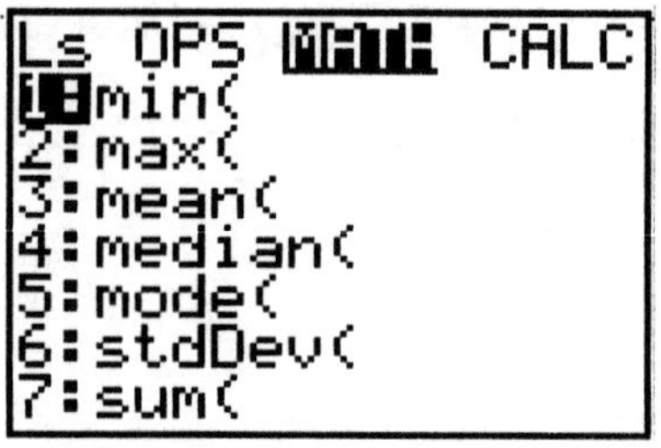

8. Select **3:mean(** by pressing **3**.

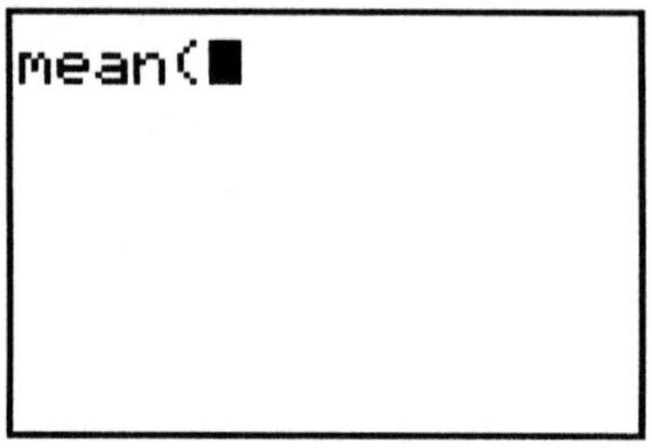

9. Press [2nd] [STAT] **4:L4** [)].

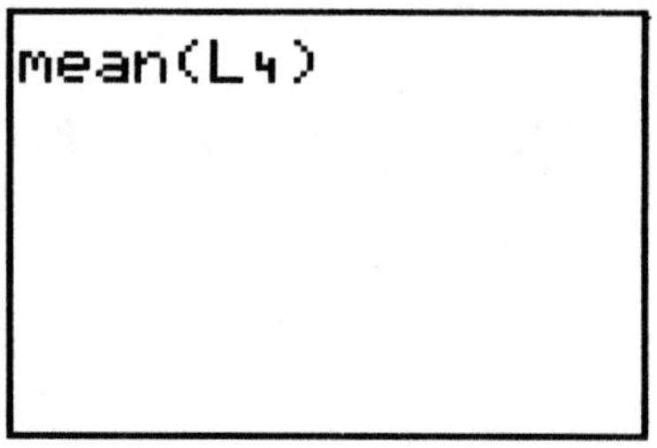

10. Press [ENTER] to calculate the mean.

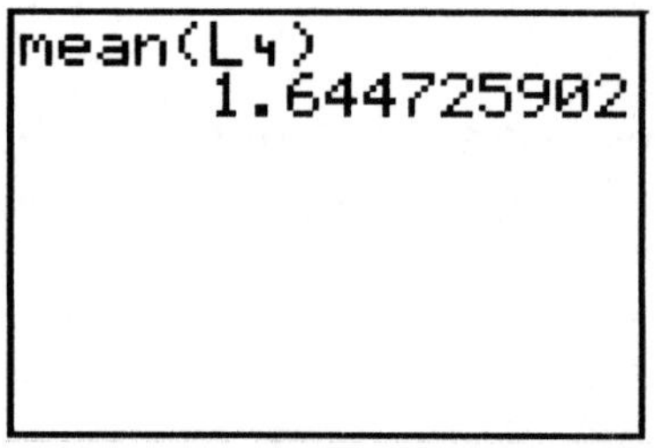

11. Repeat Steps 7 – 10 to calculate the mean of **L5**.

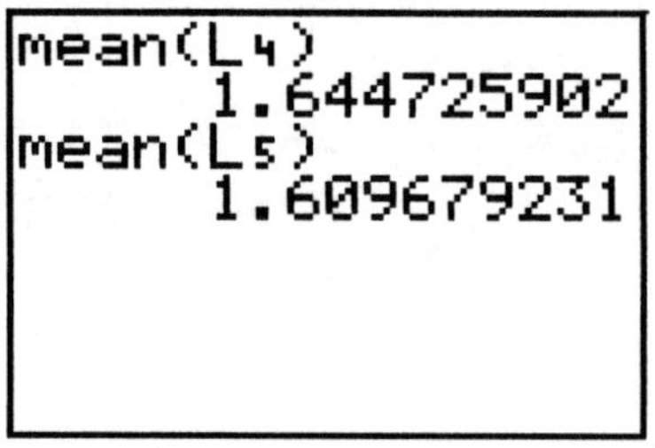

Answer questions 1 through 4 on the **Data Collection and Analysis** page.

Graphing the data: Setting up a box-and-whisker plot

1. Plot a box-and-whisker plot for the data in **L4**. Press [2nd] [PLOT]. Select **1:Plot1** by pressing **1** or [ENTER].

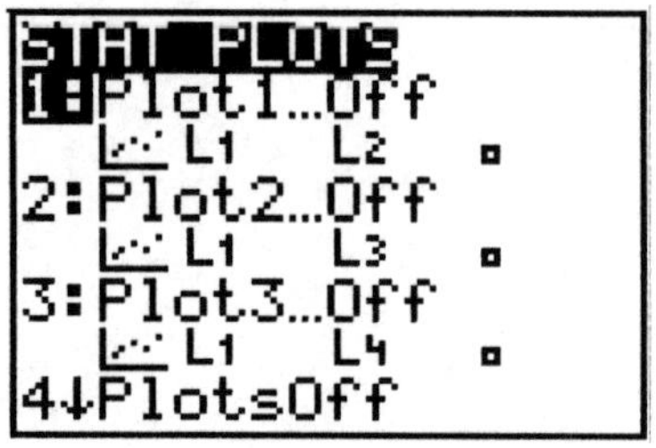

2. Set up the plot, as shown, by pressing [ENTER] [▼] [▶] [▶] [▶] [▶] [▶] [▶] [▶] [ENTER] [▼] [2nd] [STAT] **4:L4** [▼] **1** [▼] [ENTER].

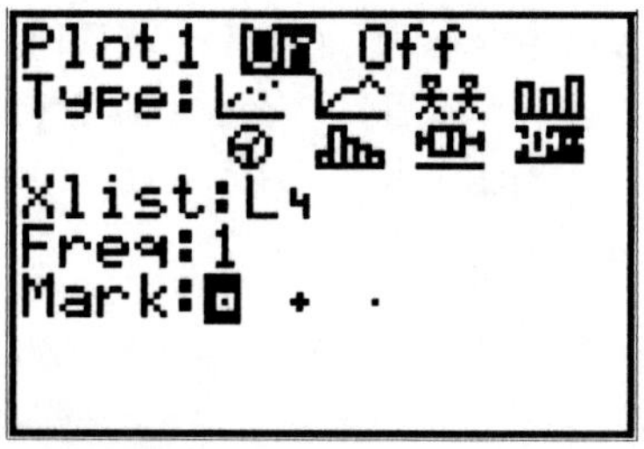

3. Press [ZOOM]. Select **7:ZoomStat** by pressing **7**.

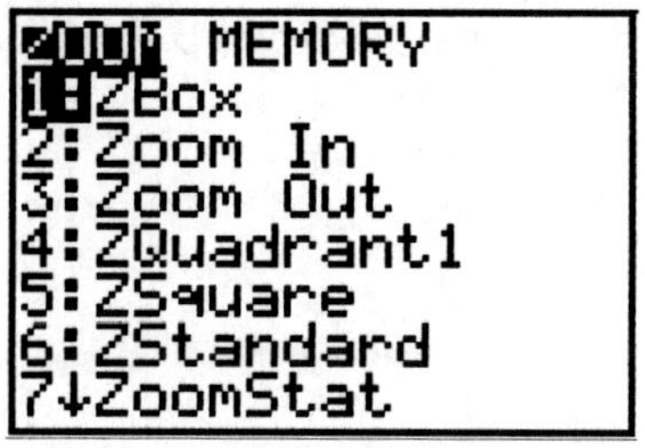

4. Press [TRACE]. Use [◀] and [▶] to see the values of the median, quartiles, and extreme values.

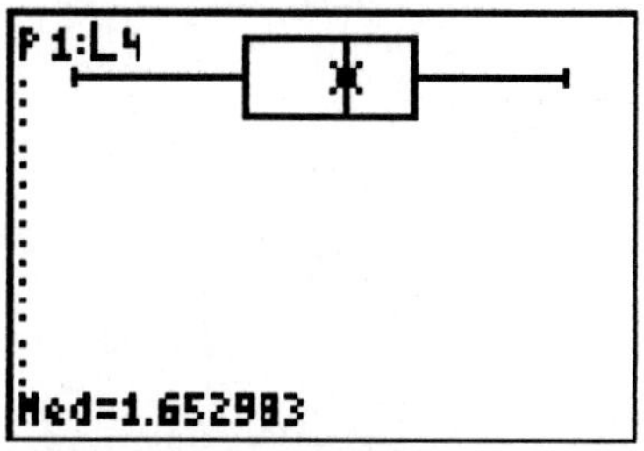

5. Plot a box-and-whisker plot for the data in **L5**. Press [2nd] [PLOT]. Select **2:Plot2** by pressing **2**.

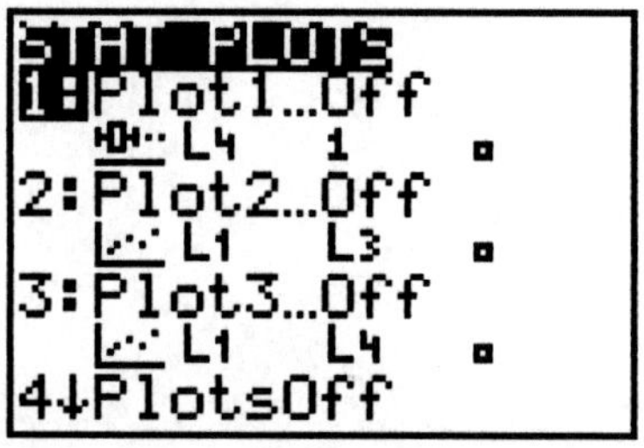

6. Set up the plot, as shown, by pressing [ENTER] [▼] [▶] [▶] [▶] [▶] [▶] [▶] [▶] [ENTER] [▼] [2nd] [STAT] **5:L5** [▼] **1** [▼] [ENTER].

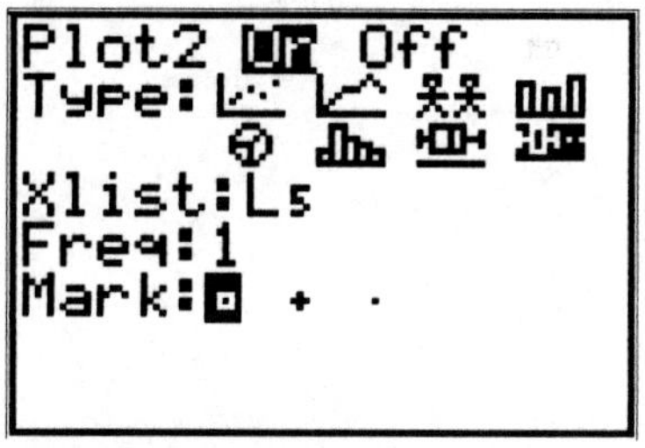

7. Press [ZOOM]. Select **7:ZoomStat** by pressing **7**.

8. Press [TRACE]. Press [▾] to move to Plot2. Use [◂] and [▸] to see the values of the median, quartiles, and extreme values.

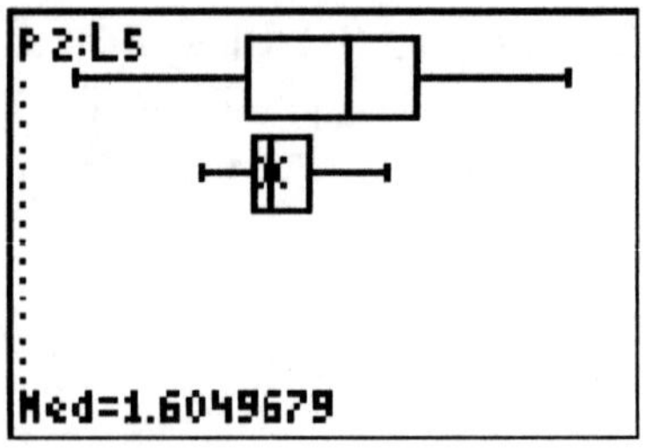

Answer questions 5 and 6 on the **Data Collection and Analysis** page.

Data Collection and Analysis

Name ______________________

Date ______________________

Activity 2: Follow the Golden Rule

Collecting the data

Record your data in the table below. You may use inches or centimeters.

Distance from head to navel	Distance from navel to floor	Distance from head to floor	Distance from shoulder to elbow	Distance from elbow to tip of hand	Distance from shoulder to tip of hand

Analyzing the data

1. What is the mean of the data for the list containing the ratio for the head to navel and navel to floor data?

   ____

2. Is the number that you entered in number 1 close to the Golden Ratio? Explain why the number might be different from the Golden Ratio.

   ____

   ____

3. If you used more students in your data collection, would you expect your value to be closer to the Golden Ratio? Why or why not?

____

____

4. Follow the directions in the **Entering the data in the TI-73** section for the shoulder to elbow, elbow to tip of hand, and shoulder to tip of hand data. What is the mean, to three decimal places, of the ratios between the shoulder to elbow measurements and the elbow to hand measurements? Is this value close to the Golden Ratio?

____

____

5. What are the lower quartile Q_1, the median, the upper quartile Q_3, and the two extreme values of the head/navel data?

Lower quartile Q_1: __________ Upper quartile Q_3: __________

Median: __________ Lower extreme: __________

Upper extreme: __________

6. What are the lower quartile Q_1, the median, the upper quartile Q_3, and the two extreme values of the navel/floor data?

Lower quartile Q_1: __________ Upper quartile Q_3: __________

Median: __________ Lower extreme: __________

Upper extreme: __________

Extensions

- Collect data on the distance from your chin to the point between your eyes and from your chin to your hairline. Set up a proportion to determine if the ratios form a Golden Ratio.
- Collect some leaves and research to find which ratios form a Golden Ratio. Determine if your leaves contain Golden Ratios. Find other species that contain the Golden Ratio, such as pine cones and the family tree of the drone bee.

Teacher Notes

Activity 2

Follow the Golden Rule

Objectives

- To use technology to find ratios
- To use technology to find measures of central tendency
- To use technology to plot a box-and-whisker plot

Materials

- TI-73 graphing device
- Metric tape measure (meter stick)

Answers to Data Collection and Analysis questions

Collecting the data

Sample data, in centimeters:

Distance from head to navel	Distance from navel to floor	Distance from head to floor	Distance from shoulder to elbow	Distance from elbow to tip of hand	Distance from shoulder to tip of hand
63	100	163	27.5	46	73.5
57	94	151	27	43	70
58.5	95	153.5	26	42	68
58	96	154	27	41.5	68.5
59	100	159	27	45	72
62	109	171	32	44	76
60	100	160	26	42	68
56.5	96	152.5	25	42	67
62	107	169	22	44	66
63	98	161	26	40.5	66.5
68	100	168	25	43	68
63	104	167	27	45	72
57	93	150	23	40	63
62	107	169	27	42	69
69	103	172	24	41	65
65	101	166	26	43.5	69.5
59	98	157	22	39	61
66	119	185	30	49	79

Analyzing the data

1. What is the mean of the data for the list containing the ratio for the head to navel and navel to floor data?

   The mean of the data is 1.645.

2. Is the number that you entered in number 1 close to the Golden Ratio? Explain why the number might be different from the Golden Ratio.

   Yes. Answers may vary. The measurements could be inaccurate.

3. If you used more students in your data collection, would you expect your value to be closer to the Golden Ratio? Why or why not?

   Yes. This would minimize the effect of any outliers.

4. Follow the directions in the **Entering the data in the TI-73** section for the shoulder to elbow, elbow to tip of hand, and shoulder to tip of hand data. What is the mean, to three decimal places, of the ratios between the shoulder to elbow measurements and the elbow to hand measurements? Is this value close to the Golden Ratio?

   The mean of the data is 1.655. Yes.

5. What are the lower quartile Q_1, the median, the upper quartile Q_3, and the two extreme values of the head/navel data?

   Lower quartile Q_1: *1.587* Upper quartile Q_3: *1.699* Median: *1.653*

   Lower extreme: *1.471* Upper extreme: *1.803*

6. What are the lower quartile Q_1, the median, the upper quartile Q_3, and the two extreme values of the navel/floor data?

   Lower quartile Q_1: *1.589* Upper quartile Q_3: *1.630* Median: *1.605*

   Lower extreme: *1.555* Upper extreme: *1.680*

Answers to Extensions questions

♦ Collect data on the distance from your chin to the point between your eyes and from your chin to your hairline. Set up a proportion to determine if the ratios form a Golden Ratio.

   Answers may vary.

♦ Collect some leaves and research to find which ratios form a Golden Ratio. Determine if your leaves contain Golden Ratios. Find other species that contain the Golden Ratio such as pinecones and the family tree of the drone bee.

   This could be assigned as a student research project.

EXPLORATIONS

Activity 3

Watching Your Weight

Objectives

- To find the *y* value of a function, given the *x* value
- To use technology to find a best fit line
- To use technology to plot a set of ordered pairs

Materials

- TI-73 graphing device
- Bathroom scale, kitchen scale, or small scale
- Block(s) equal in height to the height of the scale
- Wooden plank 2 – 4 cm thick by 25 – 30 cm wide and 120 – 140 cm long (or meter stick or ruler), one per group
- Textbooks that weigh at least 10 – 14 kilograms or 25 – 30 pounds (or a bathroom size paper cup and at least 50 pennies), one set of weights per group
- Meter stick or tape measure, one per group

Introduction

There is a toy that children used to play on called a seesaw. It is shown in the illustration to the right. Many playgrounds have removed them for safety reasons. The seesaw was a board that was hinged on a bar. When one child pushed off the ground and went up, the child on the opposite end went down. Children loved to go up and down on the seesaw. If the children were of unequal weight, it became a problem, since the heavier child would weigh down the lighter child. This problem could be solved if one of the two children moved closer to the center of the board.

Problem

How does moving a weight along a board affect the downward force on the board? How might a heavier child at one end, or a lighter child at the other end, balance a seesaw when their weights are different?

Collecting the data

To solve this problem, set up a board on a scale and add some books as shown in the illustration to the right. The board represents a portion of the seesaw. The block represents the triangular fulcrum in the diagram of the seesaw shown previously. The books are similar to the child who is sliding back and forth, trying to balance the lighter child.

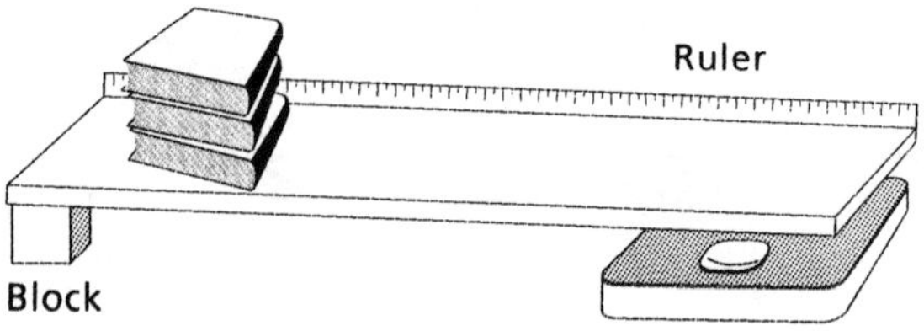

The block is used as the triangular fulcrum so that you can measure the apparent weight at the end of the board as the books slide up and down the seesaw.

1. Each group of students should obtain a scale, a wooden plank, and blocks from your teacher. Position the scale and the block(s) far enough apart so that the wooden plank is supported on one end by the scale and the other end by the block(s). Starting at the edge of the scale, place marks at 10-centimeter increments on the wooden plank. See the diagram above.

   Note: *If you have a small scale, you will use this setup with a meter stick. If you have a kitchen scale, you will use this setup with a ruler.*

2. If you are using a wooden plank, obtain textbooks that weigh at least 10 – 14 kilograms (or 25 – 30 pounds) to use for the weight. (If you are using a small scale, use a paper cup with pennies or marbles for the weight.)

3. Place the books on the wooden plank (or cup of pennies on the meter stick or ruler) at the edge of the scale. Record the weight shown on the scale for 0 centimeters in the table on the **Data Collection and Analysis** page.

4. Move the books a distance of 10 centimeters away from the scale. Record the weight shown on the scale in the table on the **Data Collection and Analysis** page.

5. Move the books a distance of 20 centimeters from the scale. Record the weight shown on the scale in the table on the **Data Collection and Analysis** page.

6. Continue to move the books away from the scale in 10-centimeter increments and record the weights.

   Note: *If you are using a meterstick, place the cup 3 cm from the scale and continue to move the cup in 3 cm increments. If you are using a ruler, place the cup 1 cm from the scale and continue to move the cup in 1 cm increments.*

Setting up the TI-73

Before starting your data collection, make sure that the TI-73 has the STAT PLOTS turned OFF, Y= functions turned OFF or cleared, the MODE and FORMAT set to their defaults, and the lists cleared. See the Appendix for a detailed description of the general setup steps.

Entering the data in the TI-73

1. Press [LIST].

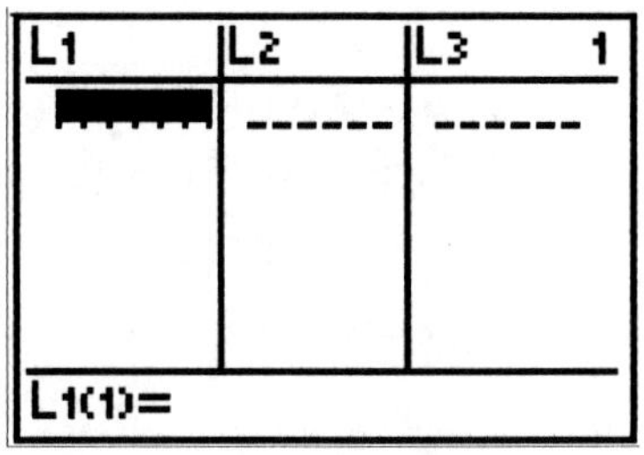

2. Enter the distance from the scale in **L1**.
3. Enter the weight shown on the scale in **L2**.

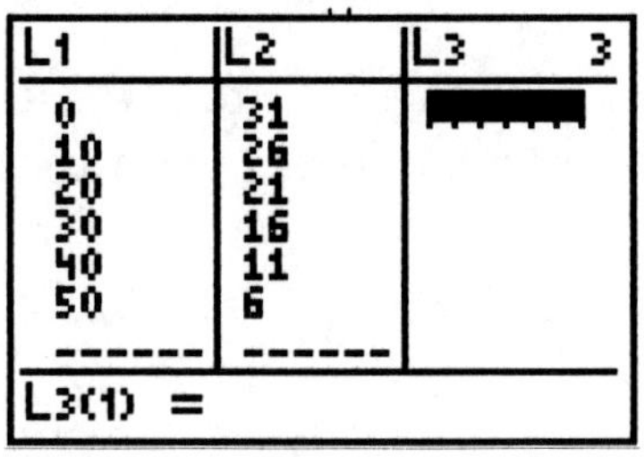

L1
L2
L3
3
0
10
20
30
40
50
31
26
21
16
11
6
L3(1) =

Setting up the window

1. Press [WINDOW] to set up the proper scale for the axes.
2. Set the **Xmin** value by identifying the minimum value in **L1**. Choose a number that is less than the minimum.

3. Set the **Xmax** value by identifying the maximum value in each list. Choose a number that is greater than the maximum. **Do Not Change the ΔX Value.** Set the **Xscl** to **10**. (Set **Xscl** to **3** if you are using a small scale or **1** if you are using a kitchen scale.)
4. Set the **Ymin** value by identifying the minimum value in **L2**. Choose a number that is less than the minimum.
5. Set the **Ymax** value by identifying the maximum value in **L2**. Choose a number that is greater than the maximum. Set the **Yscl** to **5**. (Set **Yscl** to **0.25** if you are using a small scale or a kitchen scale.)

Graphing the data: Setting up a scatter plot

1. Press [2nd] [PLOT]. Select **1:Plot1** by pressing **1** or [ENTER].

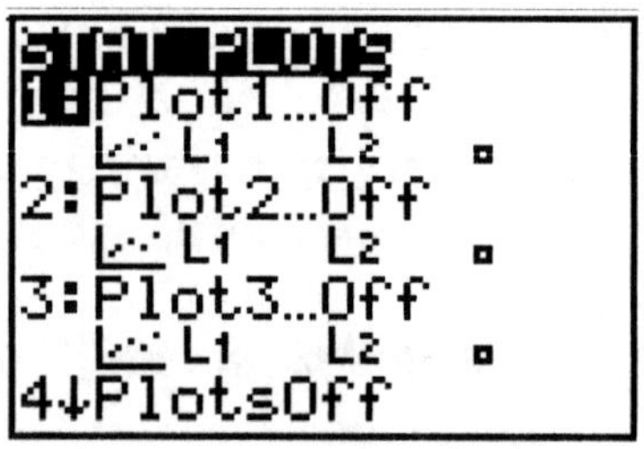

2. Set up the plot as shown by pressing [ENTER] [▼] [ENTER] [▼] [2nd] [STAT] **1:L1** [▼] [2nd] [STAT] **2:L2** [▼] [ENTER].

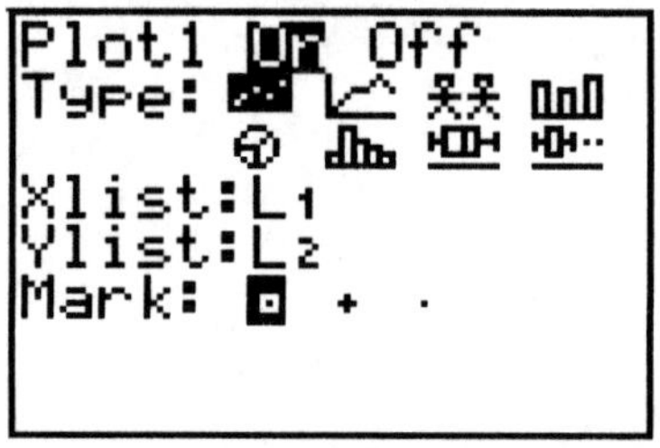

3. Press [TRACE] to see the plot. Press [◄] and [►] to move between the points.

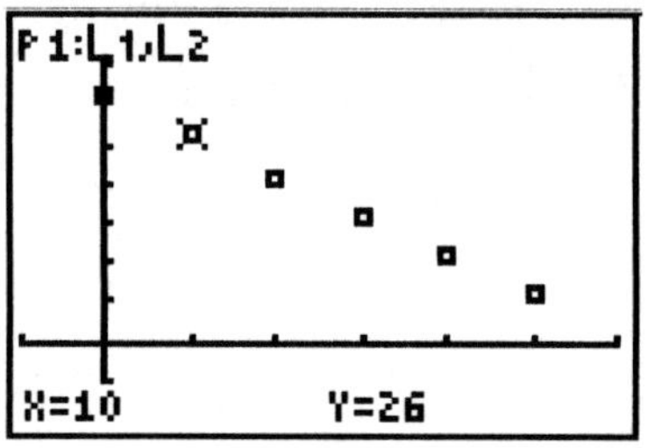

You can analyze and make predictions using the data that you collected. In order to make predictions, you must describe the data using a mathematical model. Data analysis is not an exact science, and several different methods may be used to find mathematical models. Your data may not fit any model exactly; however, the challenge is to search for a model that best fits the data. The data that you have collected should appear linear; therefore, you will find a *line of best fit* or a *trend line*. You will use two different methods to find a line of best fit or a trend line. The first method is visual and the second method uses the *linear regression* feature of the TI-73.

In the first method, use a crude but natural method for finding a line of best fit or a trend line: visually estimate the trend line. Follow the guidelines below to find the trend line.

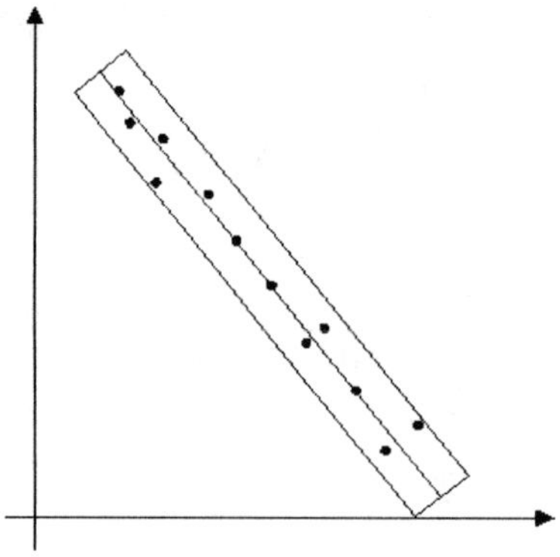

1. Find the smallest rectangle that contains all the points and shows the direction of the points.
2. Find a line that contains as many of the points as possible.
3. Find a line that divides the points equally above and below the line.

The points above or below the line should not be concentrated at one end.

In the second method, use the TI-73 to find a line of best fit. The method of finding the line of best fit employed by the TI-73 uses the formula that minimizes the sum of the squares of the residuals. A *residual* is the vertical distance between the data point and the point on the line. Look at the diagram to the right that shows the squares of the residuals. The formula used by the TI-73 finds the equation where the sum of the squares of the residuals is as small as possible. (This is referred to as the method of least-squares.)

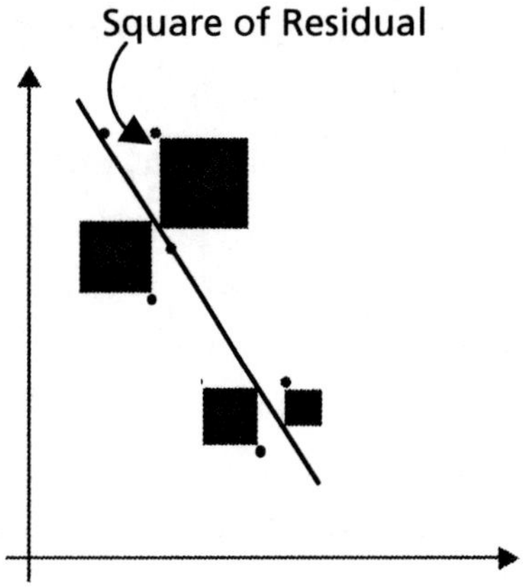

Analyzing the data

Finding a trend line (method 1)

The data appears to be linear; therefore, you can determine an equation for a trend line for the data. The equation for a line is $Y = MX + B$, where M is the *slope* and B is the y-intercept.

The *slope* of the line is defined as the change in y (Δy) divided by the change in x (Δx).

Slope = M

$$M = \Delta y/\Delta x$$

$$= \frac{(y_2 - y_1)}{(x_2 - x_1)}$$

1. Find the *slope* of the line. Press [TRACE] [▶] to move the cursor to a point on the plot. Record the x and y values shown at the bottom of the screen.

   X1 = ________ **Y1 =** ________

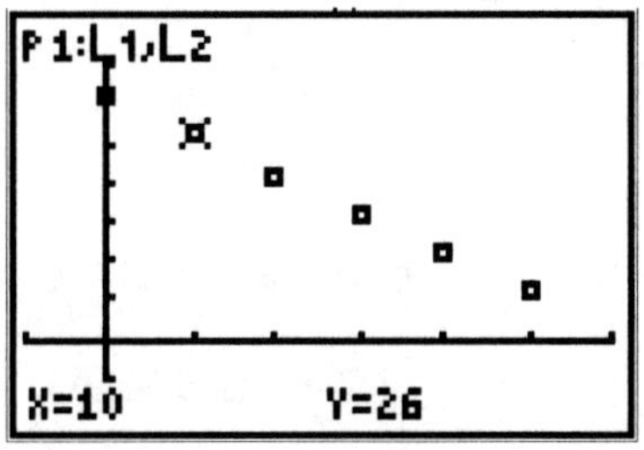

2. Press [▶] as many times as you need to find a second point on the plot. Record the x and y values shown at the bottom of the screen.

   X2 = ________ **Y2 =** ________

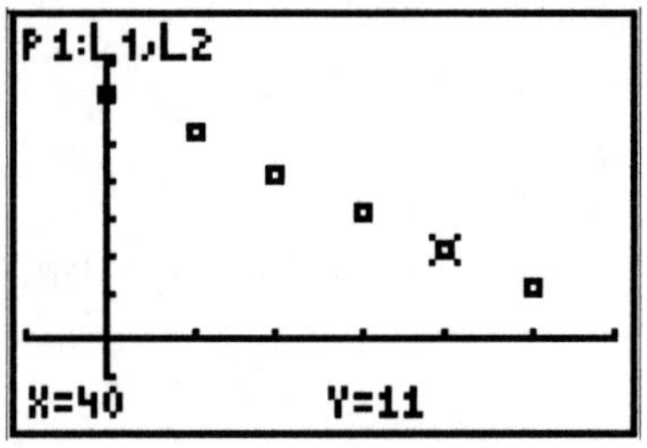

Use the following steps to calculate the *slope* of the line and to store the *slope* to **M** in the TI-73.

3. Press [2nd] [QUIT] to return to the Home screen. Press [CLEAR] to clear the Home screen.

 a. Press [(] and enter the value for **Y2**.

   ```
   (11
   ```

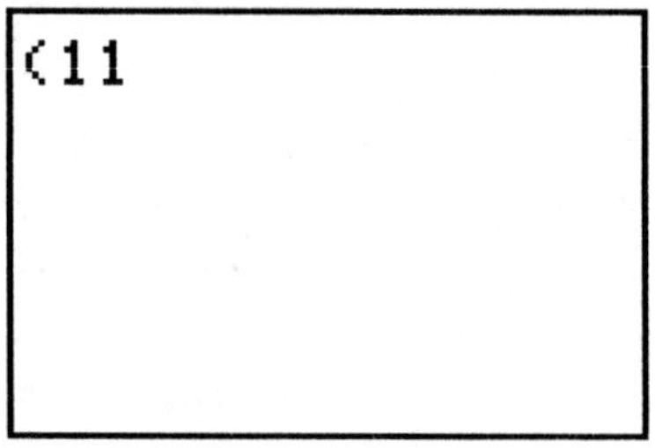

 b. Press [−] and enter the value for **Y1**.

   ```
   (11-26
   ```

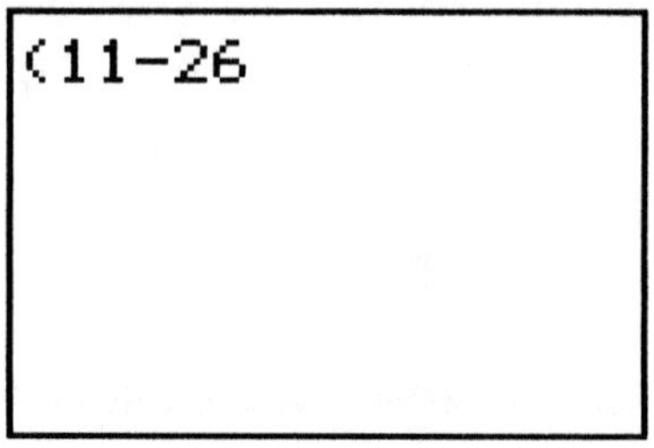

 c. Press [)] [÷] [(] and enter the value for **X2**.

   ```
   (11-26)/(40
   ```

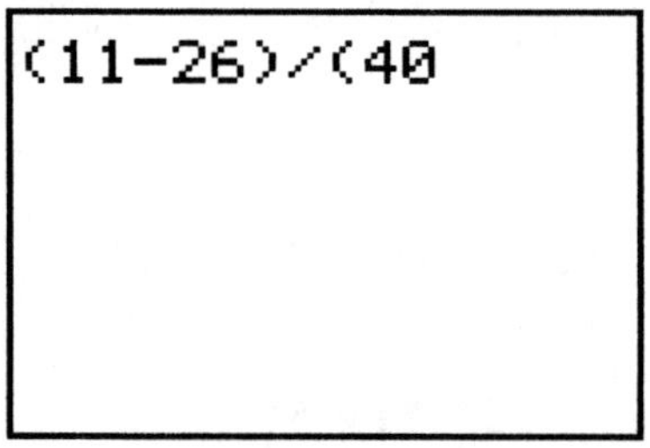

 d. Press [−] and enter the value for **X1**. Press [)].

   ```
   (11-26)/(40-10)
   ```

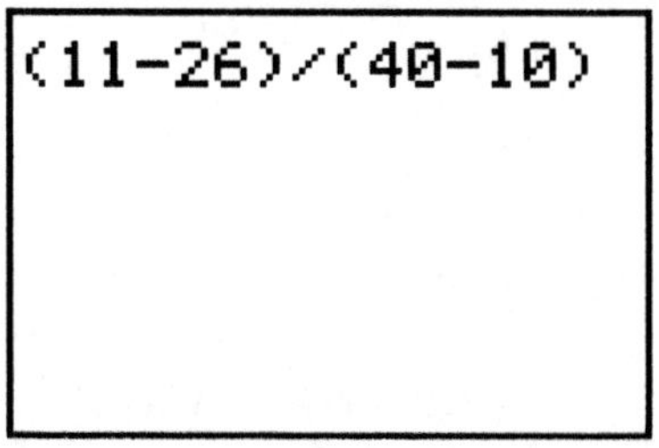

4. To store the *slope* to M:

 a. Press [STO►] [2nd] [TEXT].

   ```
   A B C D E F G H I J
   K L M N O P Q R S T
   U V W X Y Z [ ] "  _
   = ≠ > ≥ < ≤ and or
           Done
   ■
   ```

 b. Press [▼] [►] [►] [ENTER] to select M.

   ```
   A B C D E F G H I J
   K L M N O P Q R S T
   U V W X Y Z [ ] "  _
   = ≠ > ≥ < ≤ and or
           Done
   M
   ```

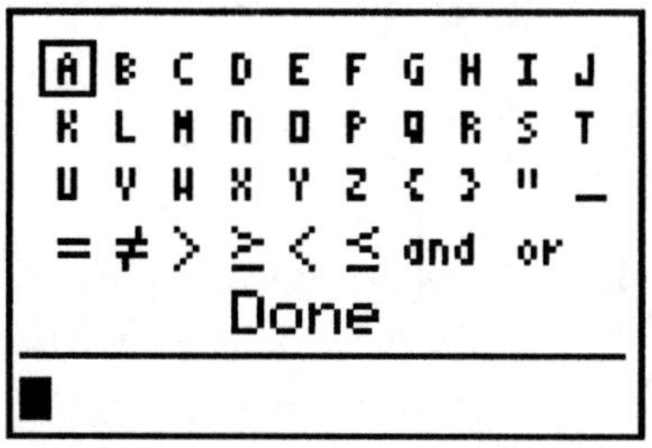

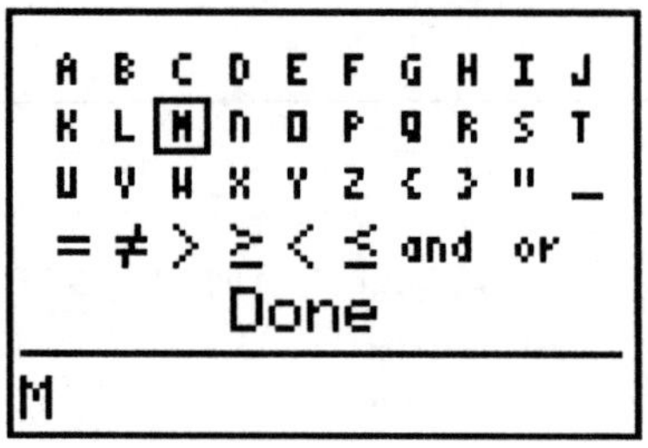

c. Press [▲] [▲] to highlight **Done**.

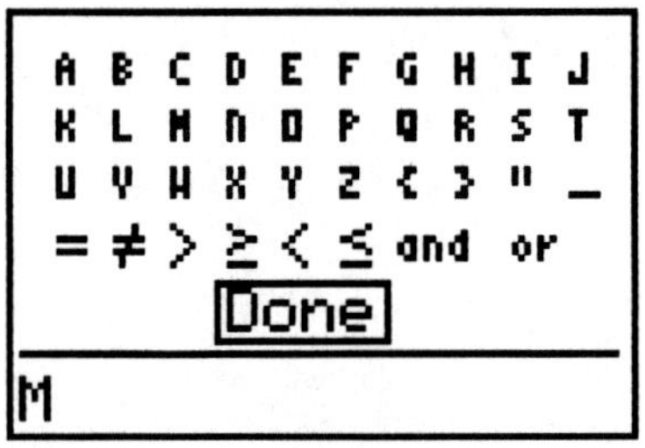

d. Press [ENTER] to exit the Text editor.

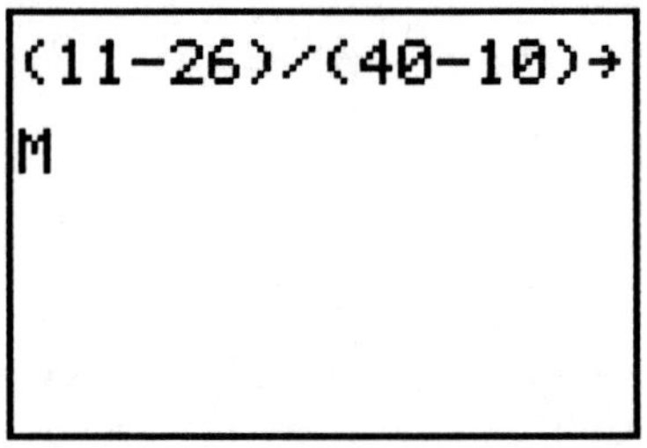

5. Press [ENTER] to calculate the *slope* and store the *slope* to **M** in the TI-73.

M= __________.

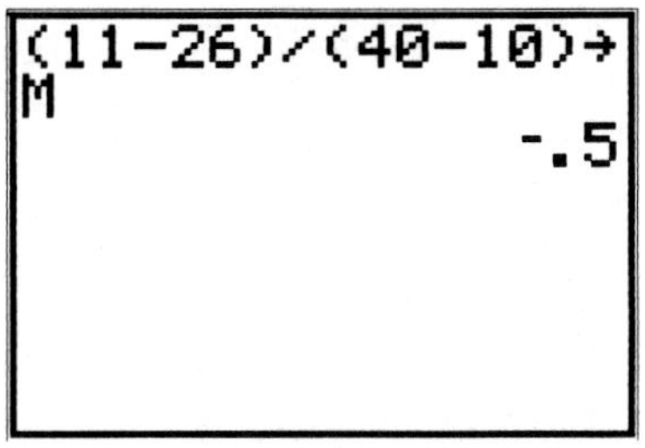

6. The *y*-intercept of a line is the point at which the line crosses the *y*-axis. The *y*-intercept of the trend line is the first value in **L2**.

B = ___________

7. Store the *y*-intercept to **B** in the TI-73.

a. Enter your *y*-intercept value, then press [STO▸].

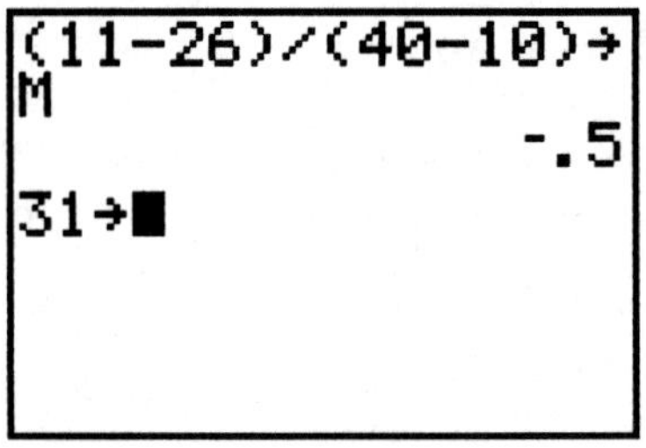

b. Press [2nd] [TEXT].

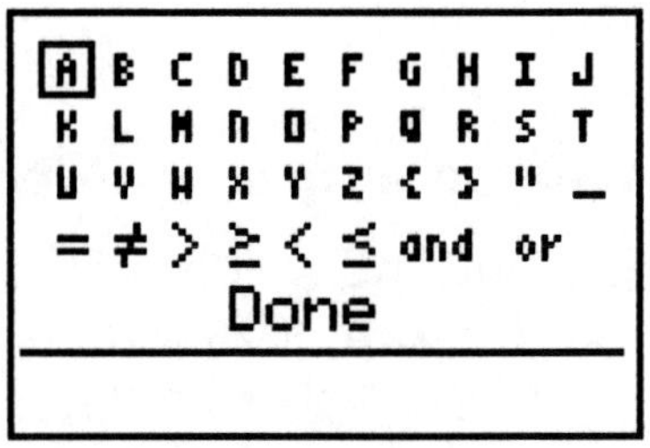

c. Press [▶] to select **B**, then press [ENTER].

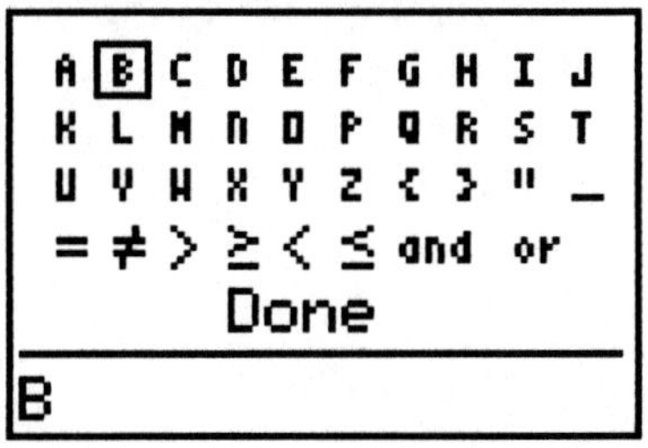

d. Press [▲] to highlight **Done**.

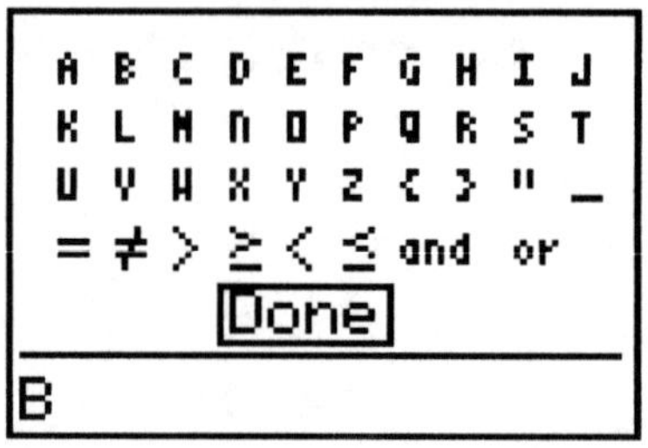

e. Press [ENTER] to exit the Text editor.

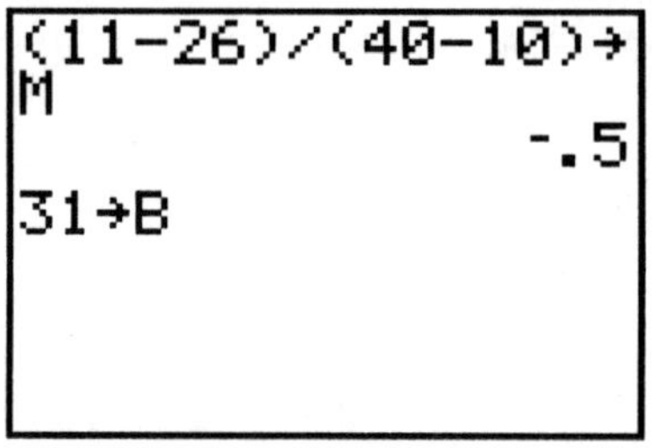

f. Press [ENTER] to store the value to **B.**

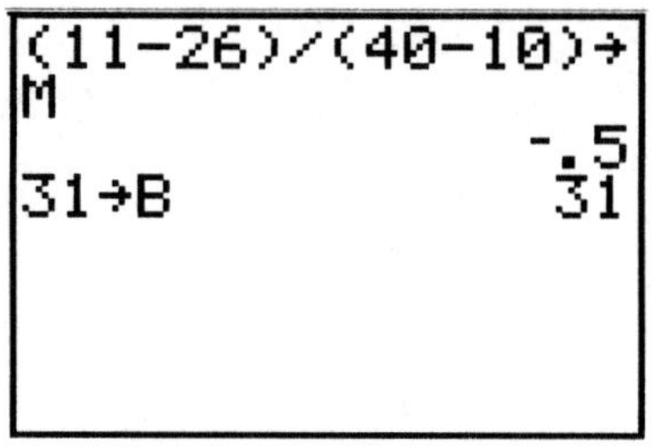

8. Enter the *slope-intercept* form of a linear equation in **Y1**.

   Press [Y=] [2nd] [TEXT] [▼] [▶] [▶] [ENTER] [▼] [▶] [ENTER] [+] [▲] [▲] [◀] [◀] [ENTER] [▲] [ENTER] to place the equation *Y* = *MX* + *B* in the Y= menu.

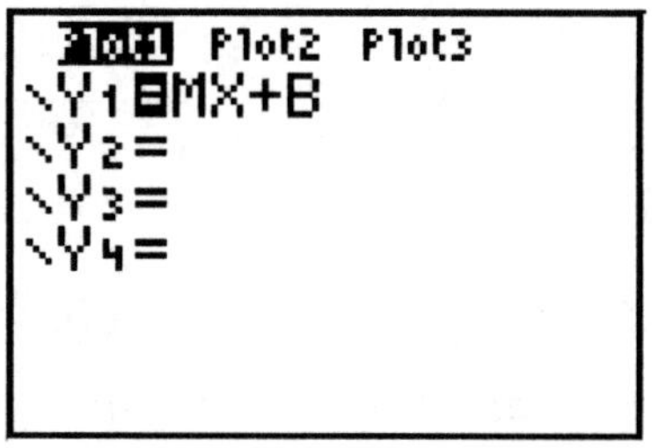

9. Press [GRAPH] to see the graph of the trend line. Record the equation on the line below.

   Equation: ______________________________

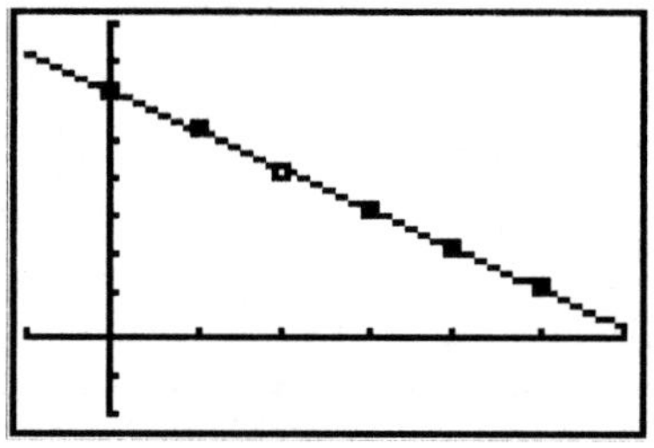

Finding the weight at different distances

1. Find the weight of the objects at a distance of 25 centimeters from the scale. Press [2nd] [TBLSET]. Press [▼] [▼] [▶] [ENTER] to set the Independent variable to **Ask**.

2. Press [2nd] [TABLE]. Enter **25.**

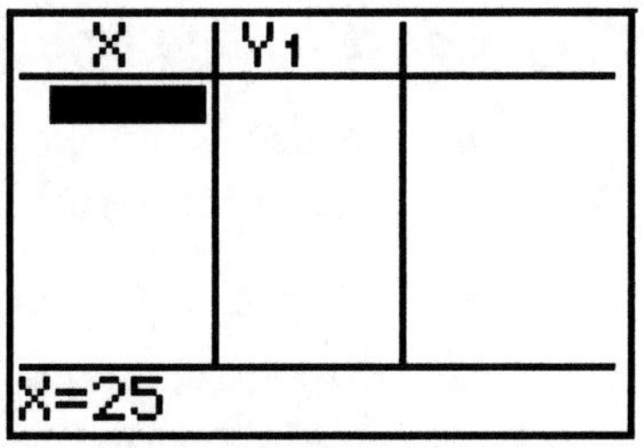

3. Press [ENTER] to see the desired weight. (The *y* value is the desired weight.)

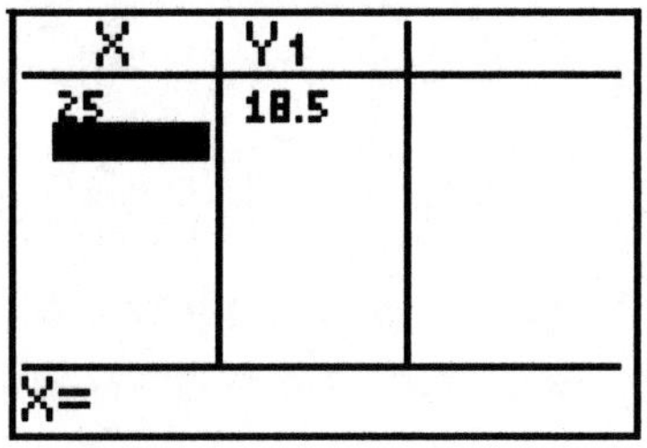

4. The [TRACE] key can be used to find the value of *y* given the value of *x*. Press [TRACE] [▼] to place the cursor on the graph of **Y1**. Enter the value of *x*, **25**.

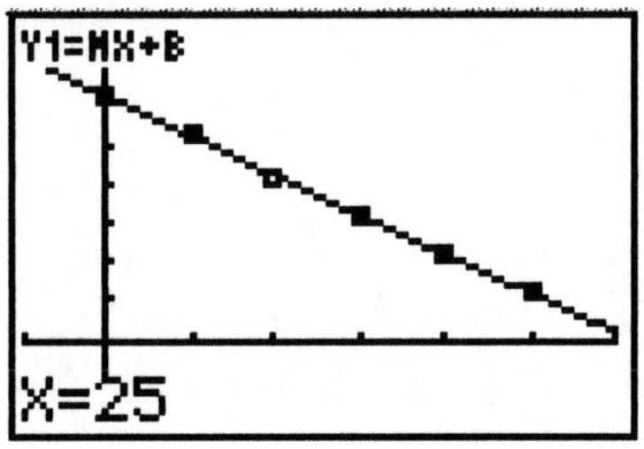

Fining the distance given the weight

1. Press [2nd] [TBLSET]. Type **0** then press [ENTER] to set the **TblStart** value. Press [▼] **1** to set the ΔTbl value. Press [▼] [ENTER] to set the independent variable back to **Auto**.

2. Press [2nd] [TABLE]. Use [▼] and [▲] to scroll the table.

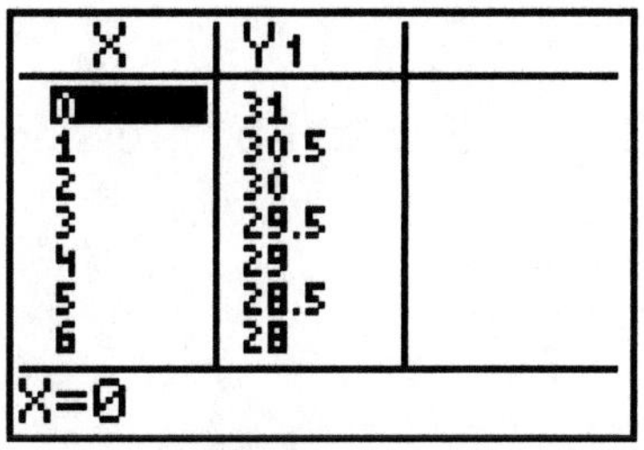

3. Find the point at which **Y1** is equal to 15.5. The *x* value is the solution to the problem.

   Note: *Your* ***Y1*** *value may not exactly equal 15.5. If this occurs, choose the value closest to 15.5.*

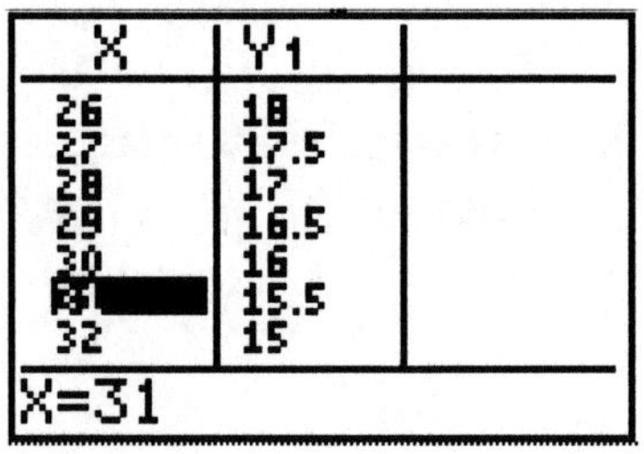

Use your equation to answer questions 1 through 6 on the **Data Collection and Analysis** page.

Finding a best fit line (method 2)

You can use the TI-73 to find the statistical line of best fit for the data. Clear the Home screen before you begin.

1. Find a linear regression equation for the data. Press [2nd] [STAT] [◄] to move the cursor to the **CALC** menu.

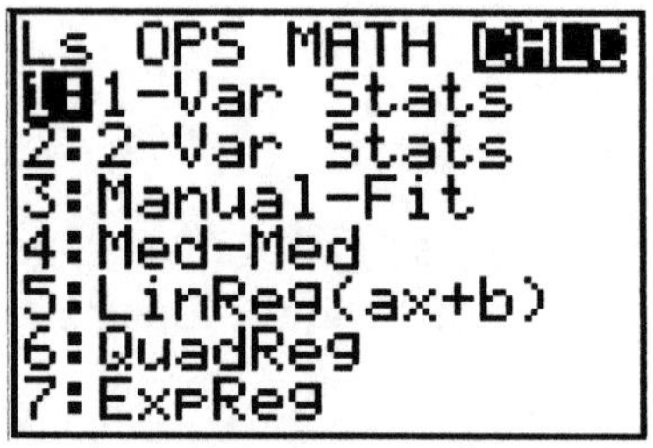

2. Select **5:LinReg(ax + b)** by pressing **5**.

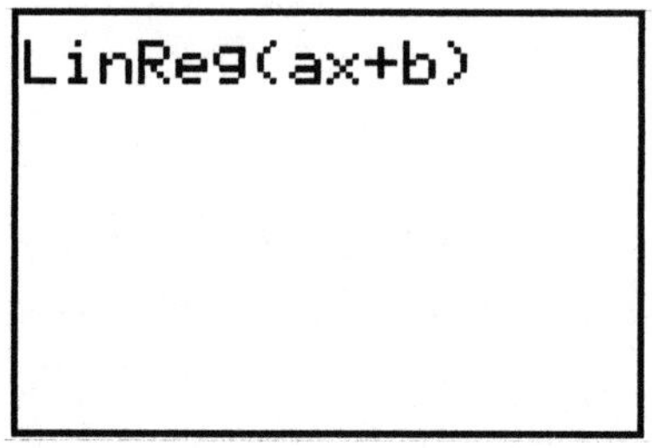

3. Enter **L_1**, **L_2**, and **Y_2**. Press [2nd] [STAT] **1:L_1** [ENTER] [,] [2nd] [STAT] **2:L_2** [,].

4. Press [2nd] [VARS]. Select **2:Y-Vars** by pressing **2**.

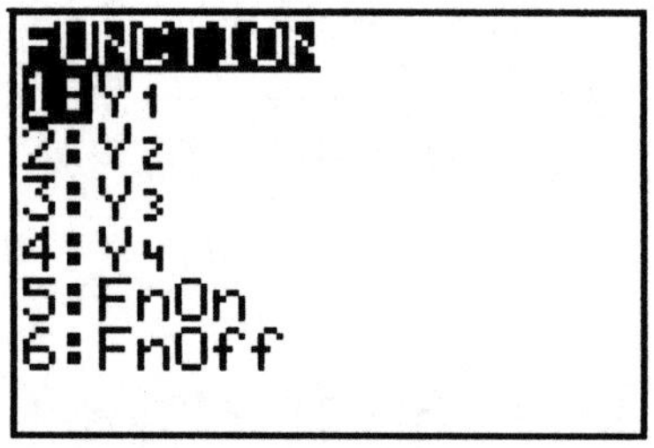

5. Select **2:Y_2** by pressing **2**.

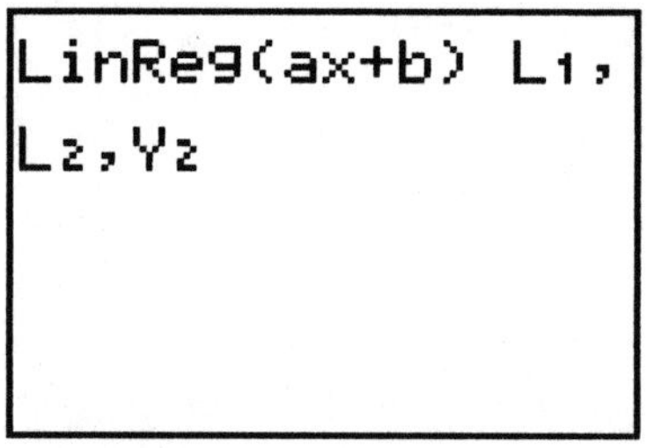

6. Press [ENTER] to calculate the equation for the best fit line. The function is pasted in **Y_2**.

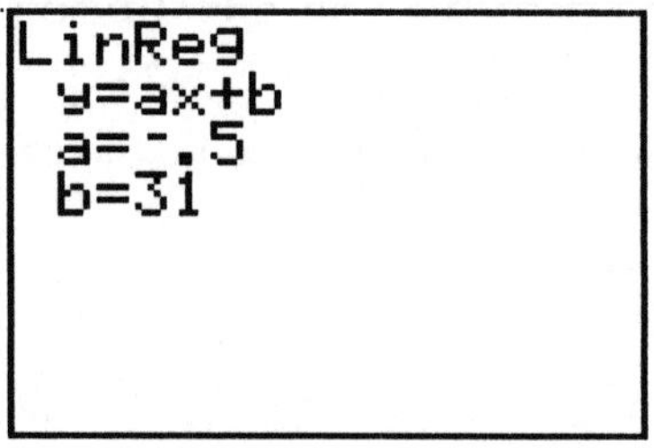

7. Press [Y=] to see the function.

   Note: *Turn OFF the equation in* **Y1**. *Press* [Y=] [◄] [ENTER].

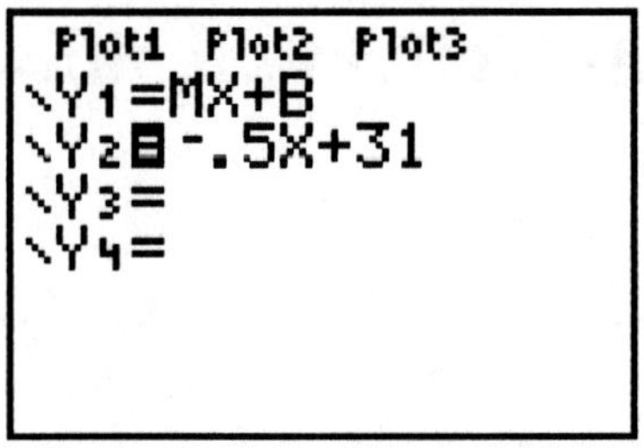

8. Press [GRAPH] to see the graph of the best fit line.

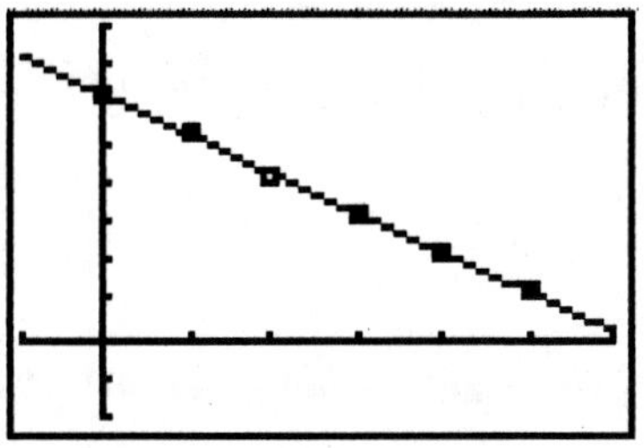

9. Repeat the **Finding the weight at different distances** and **Finding the distance given the weight** sections using the linear regression model (**Y2**).

Answer questions 7 through 10 on the **Data Collection and Analysis** page.

Data Collection and Analysis

Name ____________________

Date ____________________

Activity 3: Watching Your Weight

Collecting the data

Record your data in the appropriate column below.

Distance from scale (cm)	Weight shown on scale (__)

Analyzing the data

Use your equation from Step 9 in the **Analyzing the data: Finding a trend line (method 1)** section to answer questions 1 through 6.

1. The *slope* of the line is ______________________________ .
2. Explain what the *slope* represents.

   ____

   ____

3. The *y*-intercept of the line is ______________________________ .
4. Explain what the *y*-intercept represents.

   ____

   ____

5. Find the weight if the objects that you are using are placed at a distance of 25 centimeters from the scale. (If you used a small scale, find the weight of the objects at a distance of 13 centimeters from the scale.)

____

6. At what distance are the objects if the weight shown on the scale is one-half the original weight of the objects? ______________________________

Use the regression equation that you found in the **Analyzing the data: Finding a best fit line (method 2)** section to answer questions 7 through 10.

7. The *slope* of the regression line is ______________________________.

8. The *y*-intercept of the regression line is ______________________________.

9. Find the weight if the objects that you are using are placed at a distance of 25 centimeters from the scale. (If you used a small scale, find the weight of the objects at a distance of 13 centimeters from the scale.)

____

10. At what distance are the objects if the weight shown on the scale is one-half the original weight of the objects?

____

11. How do the values that you found in questions 1, 3, 5, and 6, using your model, compare with the values that you found in questions 7 through 10, using the regression model?

____

____

Teacher Notes

Activity 3

Watching Your Weight

Objectives

- To find the *y* value of a function, given the *x* value
- To use technology to find a best fit line
- To use technology to plot a set of ordered pairs

Materials

- TI-73 graphing device
- Bathroom scale, kitchen scale, or small scale
- Block(s) equal in height to the height of the scale
- Wooden plank 2 – 4 cm thick by 25 – 30 cm wide and 120 – 140 cm long (or meter stick or ruler), one per group
- Textbooks that weigh at least 10 – 14 kilograms or 25 – 30 pounds (or a bathroom size paper cup and at least 50 pennies), one set of weights per group
- Meter stick or tape measure, one per group

Preparation

- The wooden plank can be obtained from a lumberyard or a home improvement store. You can also use a bookshelf.
- A bathroom scale works well. You can also get a small scale from the science department in your school or use a kitchen scale.
- Make sure that students place the wooden plank at the middle of the scale.
- If you are using a bathroom scale with textbooks, make sure that the books weigh at least 10 – 14 kilograms or 25 - 30 pounds.
- If you are using small scales, use a bathroom cup with at least 50 pennies. You can also use marbles, metal washers, or any other small object with weight.

Answers to Data Collection and Analysis questions

Collecting the data

- Sample data for a bathroom scale with textbooks:

Distance from scale (cm)	Weight shown on scale (kg)
0	11.8
10	11.3
20	10.9
30	10.4
40	10.0
50	9.5
60	9.1
70	8.6
80	8.2

- Sample data for a small scale with pennies:

Distance from scale (cm)	Weight shown on scale (g)
0	4.25
3	4.00
6	3.50
9	3.00
12	2.75
15	2.25

Analyzing the data

Use your equation from number 9 in the **Analyzing the data: Finding a trend line** section to answer questions 1 through 6.

1. The *slope* of the line is ______________________________ .

Answers may vary.
For the sample data in Table 1 the slope is approximately -0.2625.
For the sample data in Table 2 the slope is approximately -1.3333.

2. Explain what the *slope* represents.

The slope represents the decrease in the number of kilograms per centimeter increase in distance from the scale.

3. The *y*-intercept of the line is ______________________________ .

The y-intercept for the sample data in Table 1 is approximately 26.
The y-intercept for the sample data in Table 2 is approximately 4.25.

4. Explain what the *y*-intercept represents.

The y-intercept represents the weight of the books at a distance of zero centimeters from the scale.

5. Find the weight if the objects that you are using are placed at a distance of 25 centimeters from the scale. (If you used a small scale, find the weight of the objects at a distance of 13 centimeters from the scale.)

The weight for the sample data in Table 1 is 19.4375 kilograms.
The weight for the sample data in Table 2 is 2.5166 grams.

6. At what distance are the objects if the weight shown on the scale is one-half the original weight of the objects?

The distance for the sample data in Table 1 is 49.52 centimeters.
The distance for the sample data in Table 2 is 15.94 centimeters.

Use the regression equation that you found in number 4 in the **Analyzing the data: Finding a best fit line** section to answer questions 7 through 10.

7. The *slope* of the regression line is ______________________________ .

For the sample data in Table 1 the slope is approximately -0.2683.
For the sample data in Table 2 the slope is approximately –0.1357.

8. The *y*-intercept of the regression line is ______________________________ .

The y-intercept for the sample data in Table 1 is approximately 25.6222.
The y-intercept for the sample data in Table 2 is approximately 4.3095.

9. Find the weight if the objects that you are using are placed at a distance of 25 centimeters from the scale. (If you used a small scale, find the weight of the objects at a distance of 13 centimeters from the scale.)

The weight for the sample data in Table 1 is 18.9138 kilograms.
The weight for the sample data in Table 2 is 2.5452 grams.

10. At what distance are the objects if the weight shown on the scale is one-half the original weight of the objects?

The distance for the sample data in Table 1 is 47.04 centimeters.
The distance for the sample data in Table 2 is 16.10 centimeters.

11. How do the values that you found in questions 1, 3, 5, and 6, using your model, compare with the values that you found in questions 7 through 10, using the regression model?

The values should be close.

EXPLORATIONS

Activity 4

The Calcumites Are Coming!

Objectives

- To model the growth of a population
- To compare ideal population growth with a population whose growth is limited
- To use technology to find an exponential and logistic regression equation
- To use technology to plot an exponential and logistic model

Materials

- TI-73 graphing device
- Calcumite cut outs, pennies, or candy

Introduction

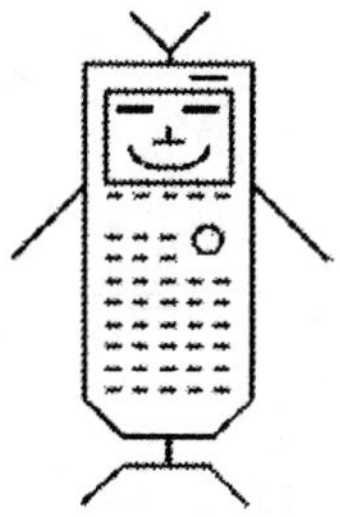

This is a Calcumite. It may look cute to the casual observer, but it is a creature that threatens to take over the world. Mutant, evolutionary cousins of your graphing calculator, these calcumites are increasing in number and will perhaps one day control the whole world! Their motto is "you push my buttons and I'll push yours." Does that sound like a friendly critter to you? Our only hope lies in understanding the population dynamics of this creature; a formidable task that will be facilitated by our TI-73, the only weapon that may save us!

Problem

Understanding population dynamics as it relates to both humans and other organisms is important in understanding the balance of nature. Ideal growth of a population refers to a population whose growth is unimpeded in any way — not by predators, disease, overcrowding, or limited resources. In practice, ideal growth cannot continue indefinitely because there are limiting factors in nature. How can we mathematically model a population's growth when it is *ideal* and when it is *limited*?

Collecting the data — Part I

In the first part of this activity, you will determine the growth rate of an *ideal* population.

1. The following page shows six generations of calcumites. Make the following assumptions about calcumites:

   - They mate when they are one year old.

- They mate once each year.
- They produce one pair (a male and a female) each time they mate.
- They never die.

2. The population for each year follows the sequence 1, 1, 2, 3, 5. To determine the next number in the sequence, carefully examine the diagram. Observe that all five pairs A, B, C, D, and E will all survive. In addition, 3 pairs, A, B, and D will have a pair of offspring, bringing the total pairs of calcumites for the next generation to 8.

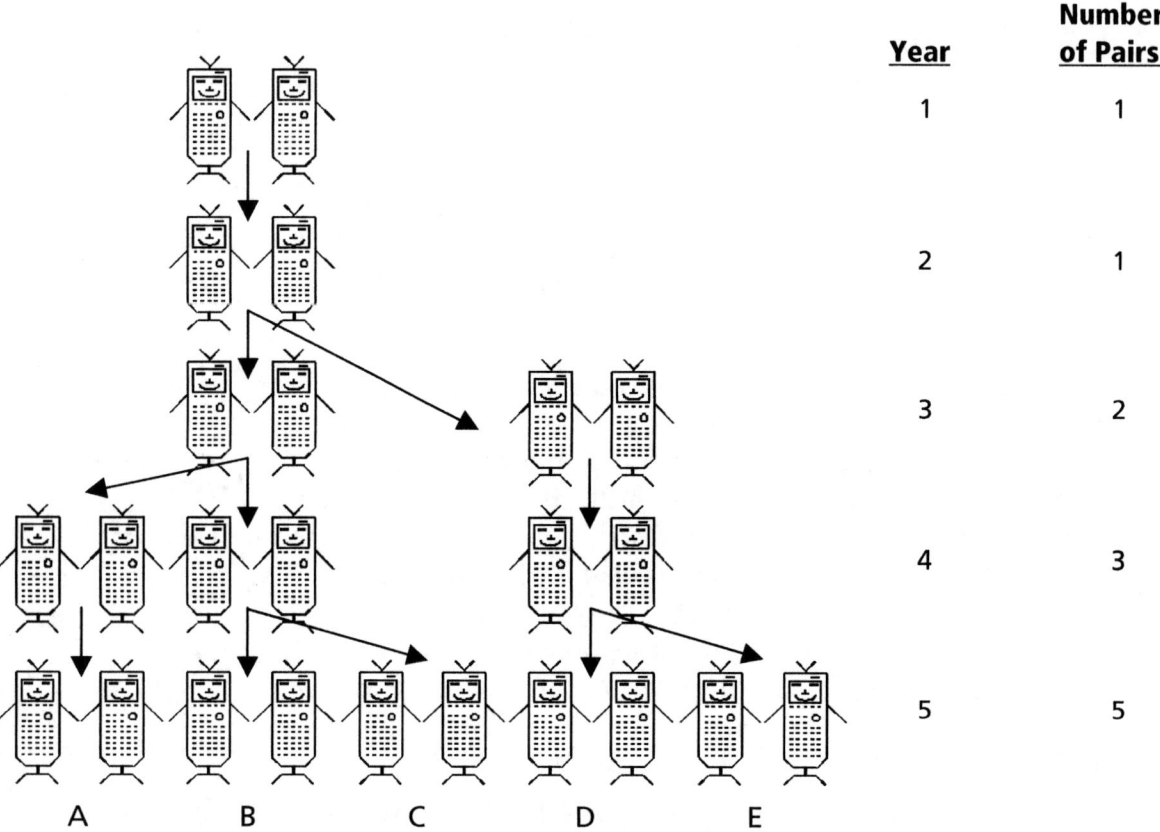

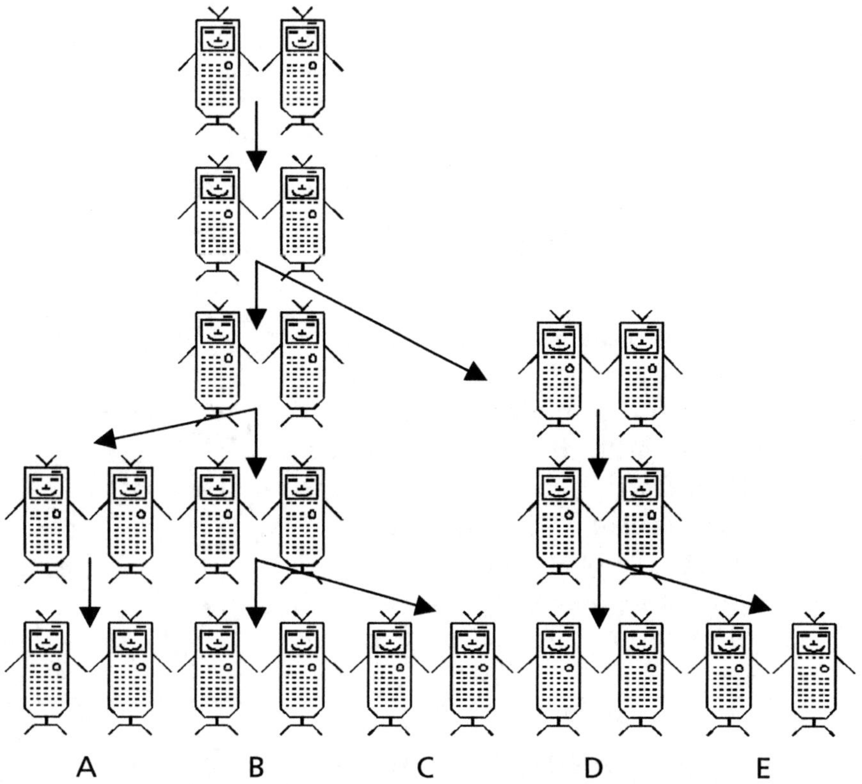

Year	Number of Pairs
1	1
2	1
3	2
4	3
5	5

Determine the relationship of numbers in this sequence. (It may be easier to work backwards.) How is the 8 related to the two integers before it? How is the 5 related to the two integers before it?

Predict the numbers after 8. Fill in the chart entitled *Ideal Population Growth — Exponential* on the **Data Collection and Analysis** page for the first 10 generations of calcumites. Check your answer with your teacher before proceeding to the next section.

Setting up the TI-73

Before starting your data collection, make sure that the TI-73 has the STAT PLOTS turned OFF, Y= functions turned OFF or cleared, the MODE and FORMAT set to their defaults, and the lists cleared. See the Appendix for a detailed description of the general setup steps.

Entering the data in the TI-73

1. Press [LIST].

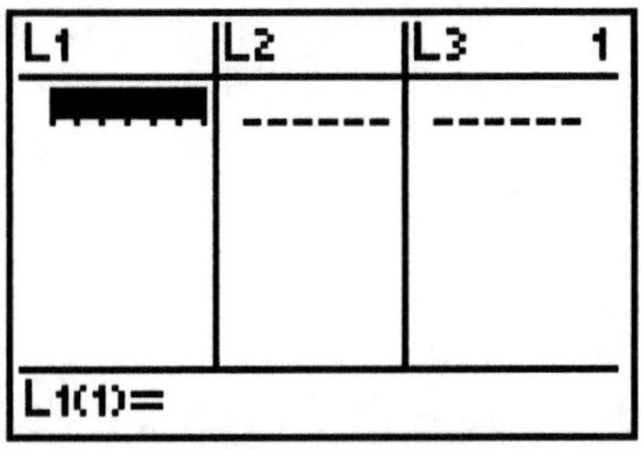

2. Enter the generation numbers 1 through 10 in **L1**.
3. Enter the number of calcumite pairs in **L2**. (Make sure that the pairs of generation numbers and calcumite pairs match in each column.)

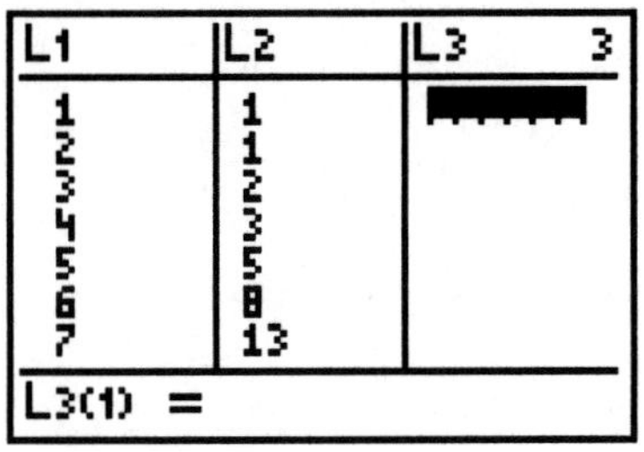

Setting up the window

1. Press [WINDOW] to set up the proper scale for the axes.
2. Set the **Xmin** value by identifying the minimum value in **L1**. Choose a number that is less than the minimum.
3. Set the **Xmax** value by identifying the maximum value in each list. Choose a number that is greater than the maximum. **Do Not Change the ΔX Value.** Set the **Xscl** to **1**.

4. Set the **Ymin** value by identifying the minimum value in **L2**. Choose a number that is less than the minimum.
5. Set the **Ymax** value by identifying the maximum value in **L2**. Choose a number that is greater than the maximum. Set the **Yscl** to **5**.

Graphing the data: Setting up a scatter plot

In order to analyze the data, you will need to set up a scatter plot and model the data by graphing the results of an exponential regression.

1. Press [2nd] [PLOT]. Select **1:Plot1** by pressing **1** or [ENTER].

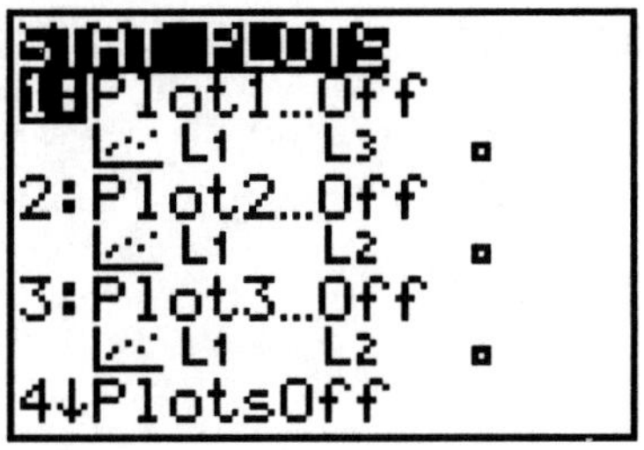

2. Set up the plot as shown by pressing [ENTER] [▼] [ENTER] [▼] [2nd] [STAT] **1:L1** [▼] [2nd] [STAT] **2:L2** [▼] [ENTER].

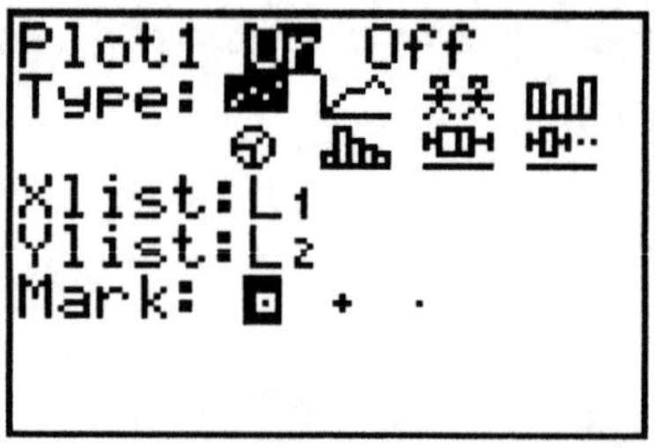

3. Press [GRAPH] to see the plot.

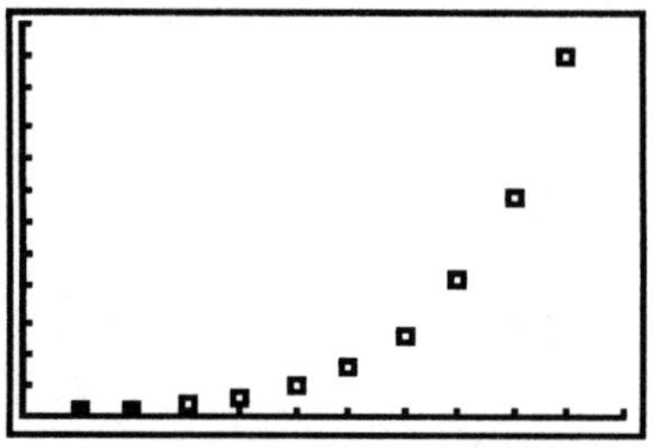

4. The graph to the right shows the rate of calcumite population growth for three intervals of the plot shown in number 3:

   2 to 3 generations

   5 to 6 generations

   8 to 9 generations

   Determine the growth rate for each of those time intervals by determining the *slope* of each of the plots, using the following formula for slope:

   $$\text{Slope} = \frac{(y_2 - y_1)}{(x_2 - x_1)}$$

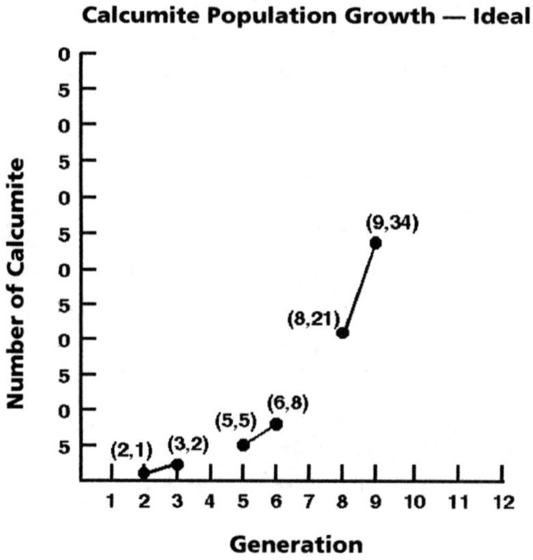

Enter the slopes on the **Data Collection and Analysis** page.

Answer Part I questions 1 and 2 on the **Data Collection and Analysis** page.

You have examined linear models in previous activities. In a linear model, the slope does not change. Therefore, a linear model would not be appropriate here. What type of regression would be a suitable model for this data?

Analyzing the data: Finding an exponential regression

1. Press [2nd] [STAT] [◄] to move the cursor to the **CALC** menu.

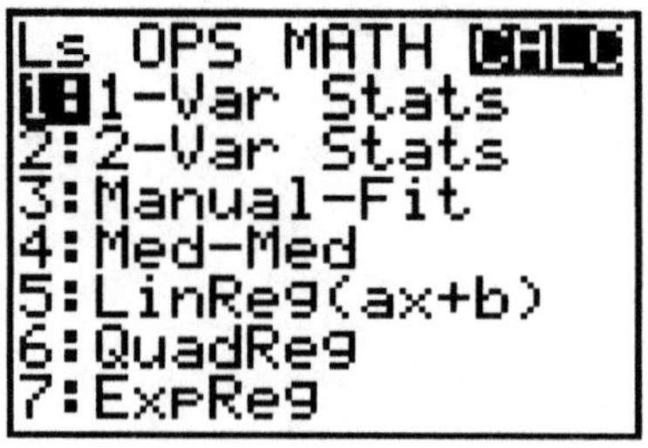

2. Select **7:ExpReg** by pressing **7**.

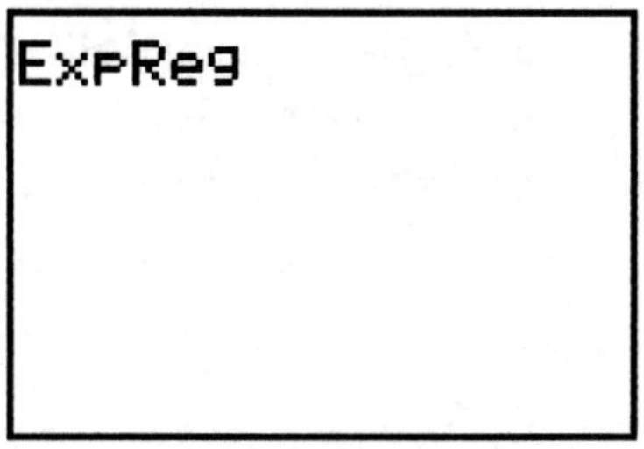

3. Press [2nd] [STAT] **1:L1** [,] [2nd] [STAT] **2:L2** [,].

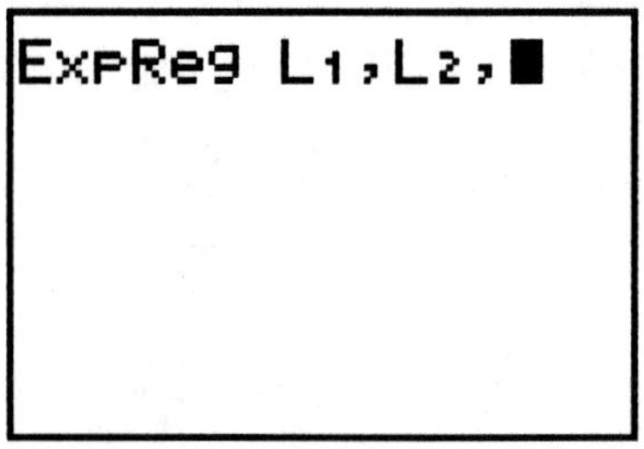

4. Press [2nd] [VARS]. Select **2:Y-Vars** by pressing **2**.

5. Select **1:Y1** by pressing **1** or [ENTER].

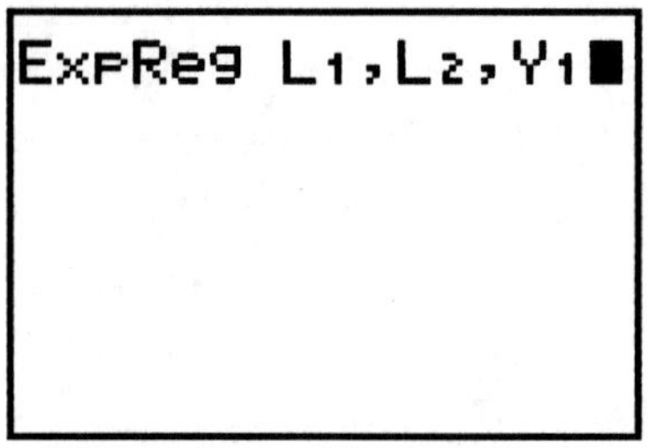

6. Press [ENTER] to calculate the exponential regression. The function is pasted into **Y1**.

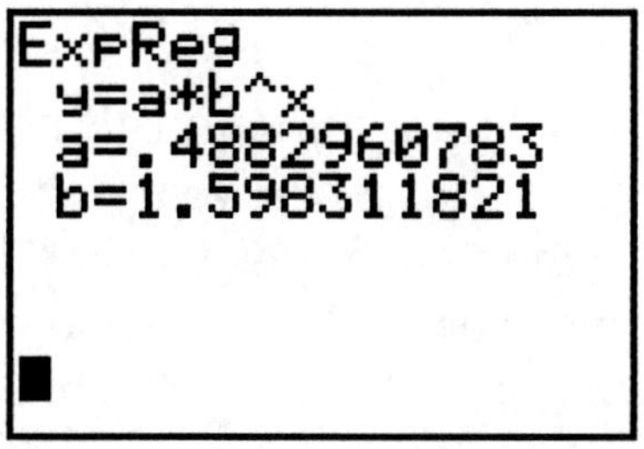

7. Press [GRAPH] to see a graph of the exponential regression model.

   The plot displayed shows the ideal calcumite population growth over a period of 10 generations.

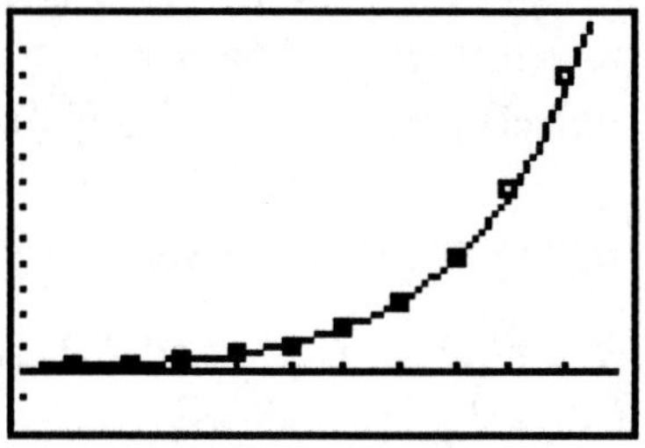

8. Examine how large the population will become after 20 generations. To do this, the window must be expanded.

   Press [WINDOW] and set the window values as shown.

   Do NOT change the ΔX value.

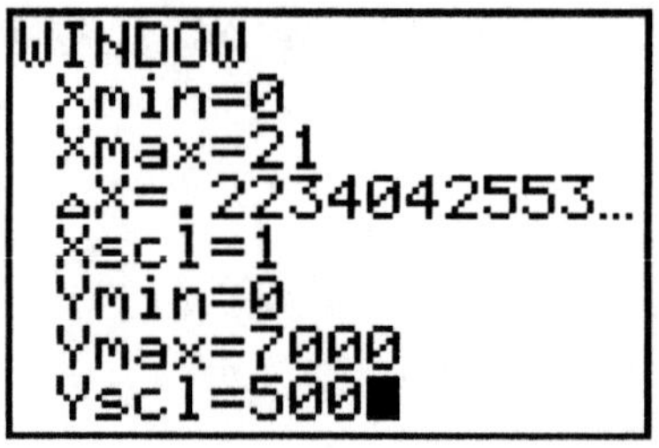

9. Press [GRAPH] to see the plot.

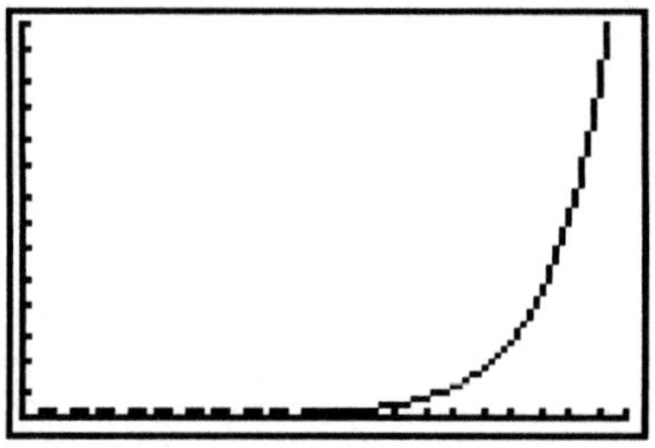

10. Find the population of calcumites after 20 generations by pressing [TRACE] [▼] to place the cursor on the graph of **Y1**.

    Type **20** and press [ENTER] to show the number of calcumite pairs (*y* value) after 20 generations of population growth.

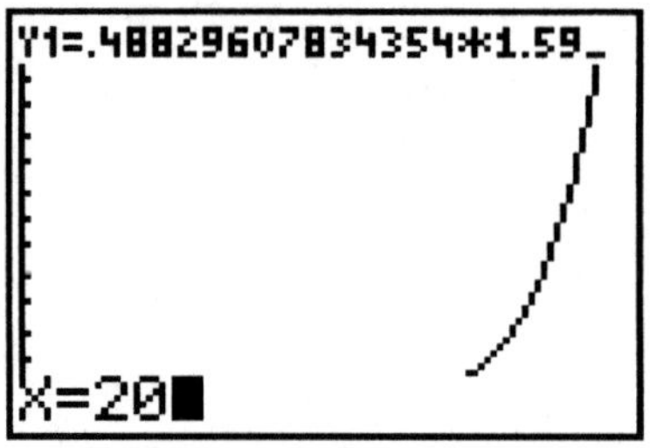

11. Record the number of pairs of calcumites after 20 generations on the **Data Collection and Analysis** page.

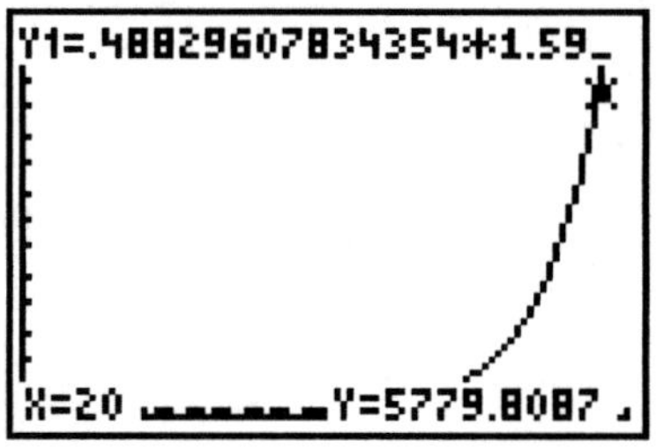

It becomes apparent that a population cannot keep rising at this rate. Keep in mind that a one-year generation time for calcumites is somewhat long compared to many other organisms. For example, the generation time of a fruit fly is about two weeks. Also, keep in mind that some organisms produce many offspring, not just one pair. For example, tapeworms may produce 600 million eggs annually, but only a few survive. There are always factors that limit the growth of a population. If this were not true, the earth would soon be covered with fruit flies, tapeworms, or our imaginary calcumites!

Collecting the data — Part II

In the second part of this activity, you will investigate the growth rate of a *limited* population.

As the calcumite population grows and becomes denser, more of them die for various reasons. Some die from disease. (They are infected with a computer-like virus that attacks calcumites.) Some calcumites are eaten by TI-73 calculators. Others perish because they cannot find enough food and resources. The following chart shows 10 generations of calcumites whose growth is *limited*. Plot the growth of this population and analyze the consequences of such limited growth.

Generation	1	2	3	4	5	6	7	8	9	10
Pair Number	1	1	2	3	5	8	10	11	12	12

Entering the data in the TI-73

1. Press [LIST].

   The generation data in **L1** is the same as in the study of ideal population growth and need not be changed.

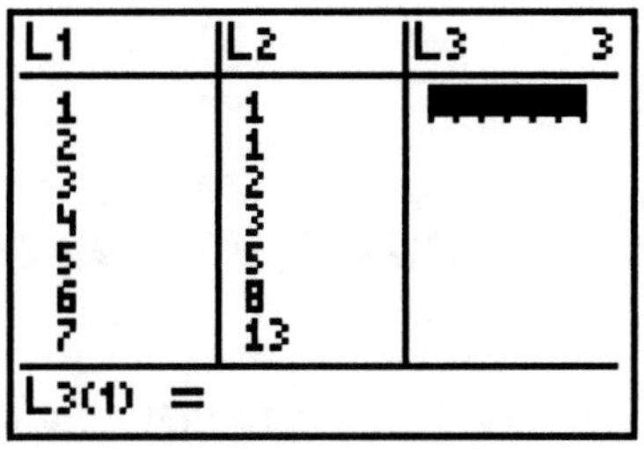

2. Clear the data from **L2** by moving the cursor to the **L2** label and pressing [CLEAR] [ENTER].

   Enter the number of calcumite pairs for limited population growth into **L2**.

   (Make sure that the pairs of generation numbers and calcumite pairs match in each column.)

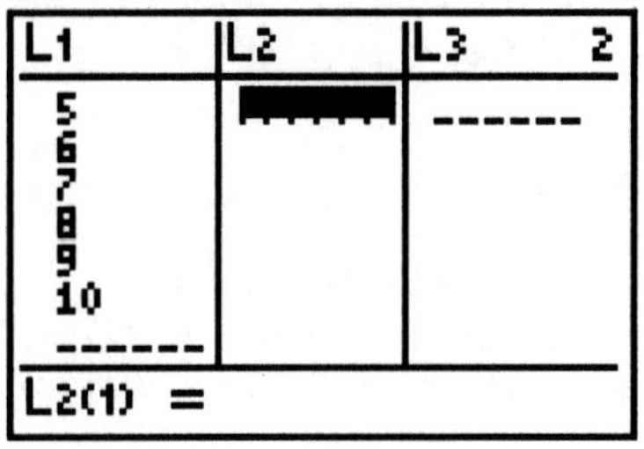

Setting up the window

1. Press [WINDOW] to set up the proper scale for the axes.
2. Set the **Xmin** value by identifying the minimum value in **L1**. Choose a number that is less than the minimum.

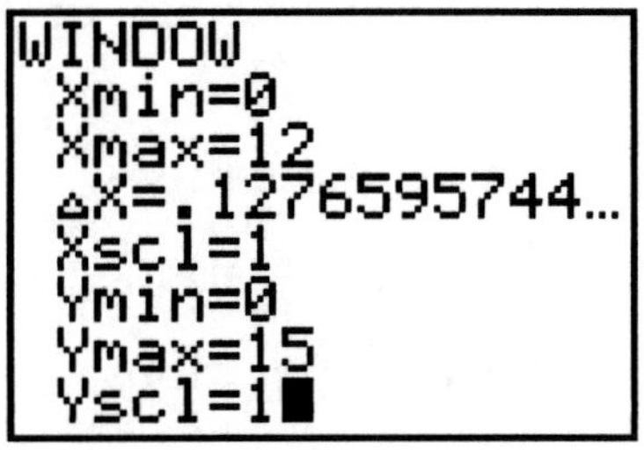

3. Set the **Xmax** value by identifying the maximum value in each list. Choose a number that is greater than the maximum. **Do Not Change the ΔX Value.** Set the **Xscl** to **1**.
4. Set the **Ymin** value by identifying the minimum value in **L2**. Choose a number that is less than the minimum.
5. Set the **Ymax** value by identifying the maximum value in **L2**. Choose a number that is greater than the maximum. Set the **Yscl** to **1**.

Graphing the data: Setting up a scatter plot

In order to analyze the data, you will need to set up a scatter plot and model the data by graphing a logistic function.

1. Press [2nd] [PLOT]. Select **1:Plot1** by pressing **1** or [ENTER].

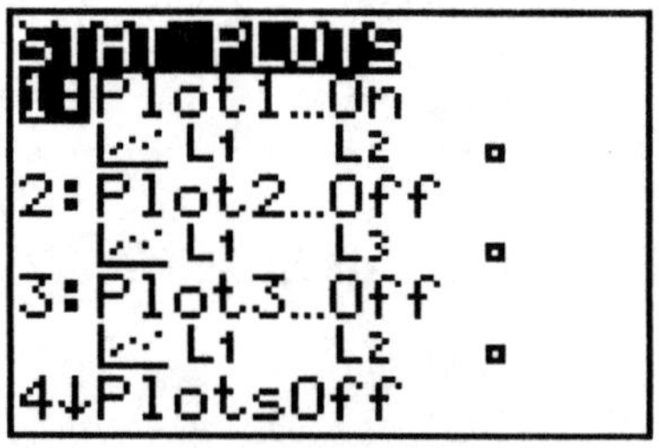

2. Set up the plot as shown by pressing [ENTER] [▼] [ENTER] [▼] [2nd] [STAT] **1:L1** [▼] [2nd] [STAT] **2:L2** [▼] [ENTER].

   Note: *This step may be unnecessary if settings were not changed after you plotted ideal population growth.*

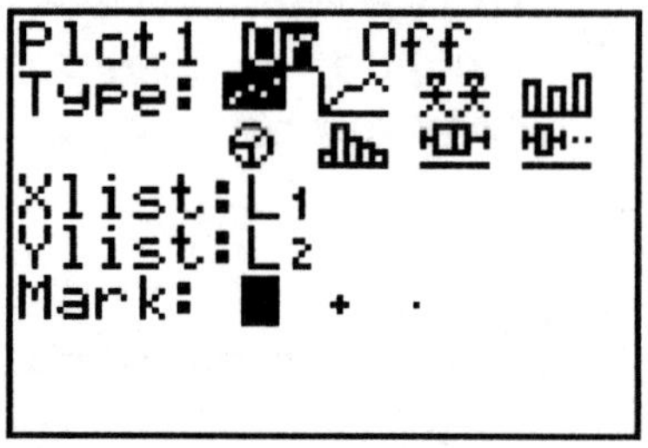

3. Press [GRAPH] to see the plot.

   Note: *The exponential regression model for your ideal population is still on the screen. You will use that model for comparison purposes.*

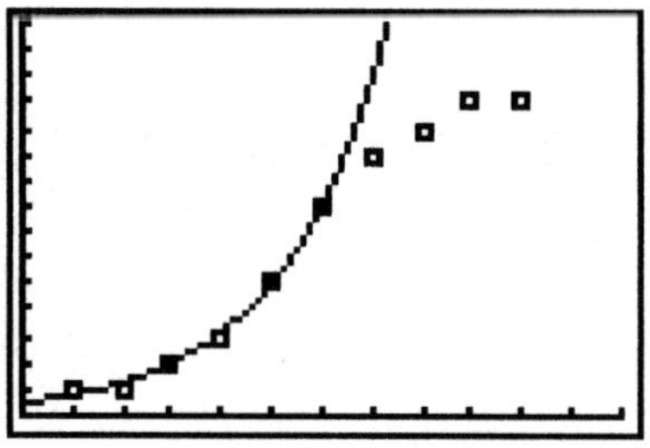

4. The graph to the right shows the rate of calcumite population growth for three intervals of the plot shown in number 3.

   2 to 3 generations

   5 to 6 generations

   8 to 9 generations

   Determine the growth rate for each of those time intervals by determining the *slope* of each of the plots, using the following formula for slope:

   $$\text{Slope} = \frac{(y_2 - y_1)}{(x_2 - x_1)}.$$

Calcumite Population Growth — Limited

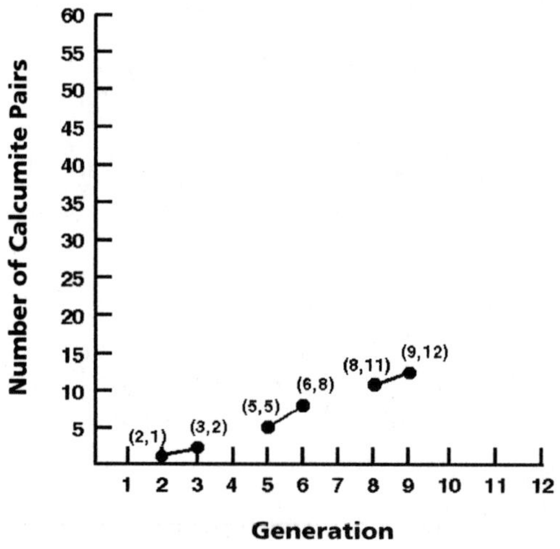

Enter the slopes on the **Data Collection and Analysis** page.

Answer Part II questions 1 and 2 on the **Data Collection and Analysis** page.

Analyzing the data: Finding a logistic regression

It is necessary to determine an appropriate regression model for this data. You know it is not linear since the slope changes. An exponential regression would not be appropriate, since the slope does not continually increase over the three intervals. What type of regression would be a suitable model for this data?

The function that best models this data is a logistic function. This is a complex equation (beyond the scope of your course), yet once plotted, you will be able to analyze it and learn how it most accurately shows population growth in nature.

Using the data provided, the equation is as follows:

$$Y = 12.6 / (1 + 61.6e^{-0.77X})$$

1. Press [Y=]. Scroll down to **Y2**.

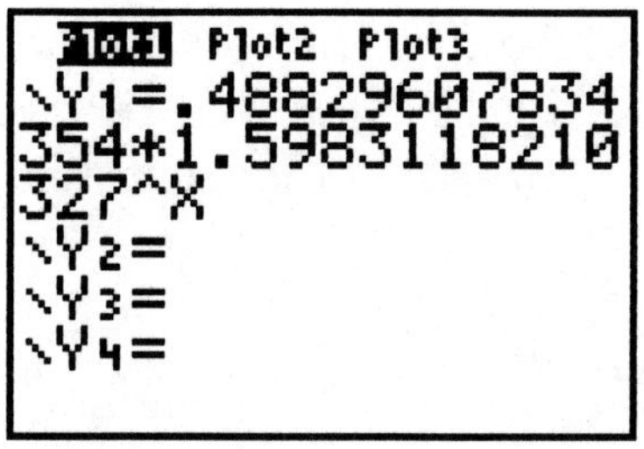

2. Type **12.6** [÷] [(] **1** [+] **61.6** [MATH] [◄] **4:e^(** [(-)] **.77** [x] [)] [)].

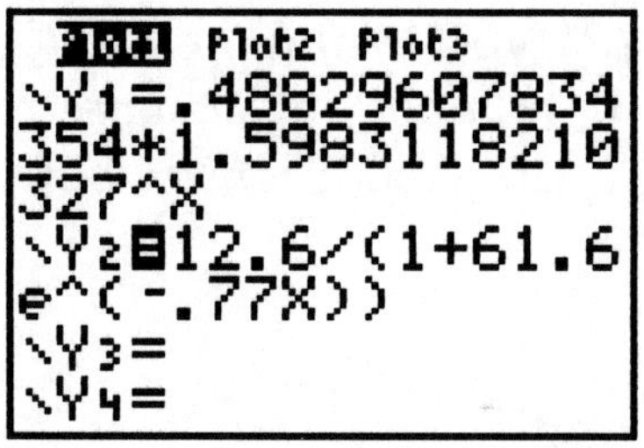

3. Press [GRAPH] to view a graph of the logistic function.

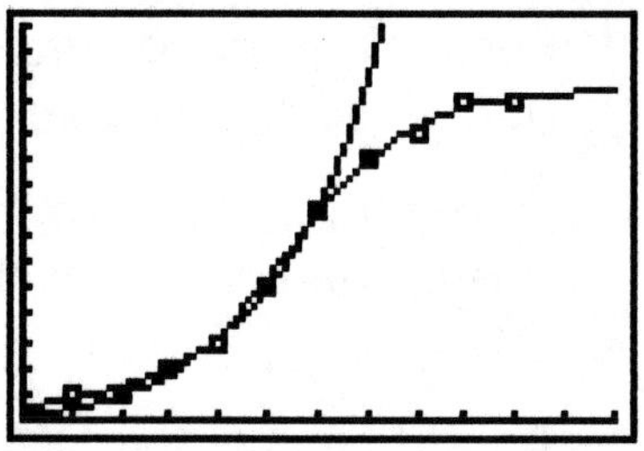

The logistic function is complex beyond the scope of your studies this year. You can, however, make a comparison between a population growing exponentially and one that is growing logistically.

It becomes obvious that the rate at which the population grows is changing; it starts slow, speeds up, and then slows down. This is different from what you observed with ideal population growth (exponential). In order to examine how large the population will become after 20 generations, it is necessary to expand the window.

4. Press [WINDOW] and set the window values as shown.

   Do Not Change the ΔX Value.

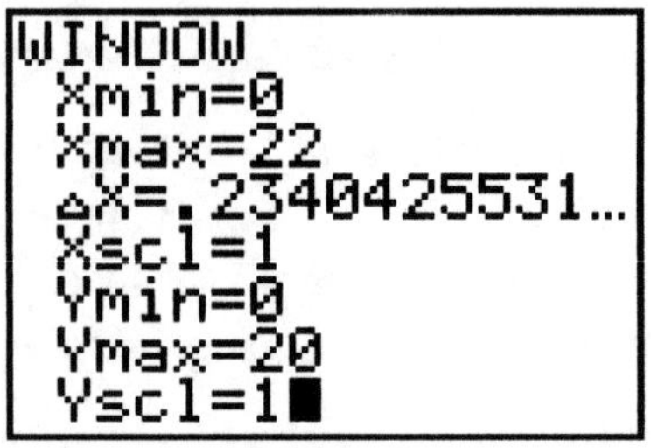

5. Press [GRAPH] to see the enlarged generation growth graph.

   Note: *This plot levels off. The population size where it levels off is referred to as the carrying capacity.*

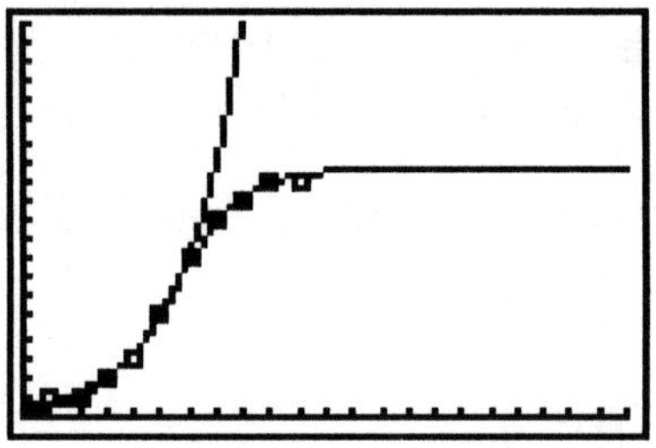

6. To find the carrying capacity, press [TRACE] [▲] so that the cursor is on the graph of the **Y2** logistic function.

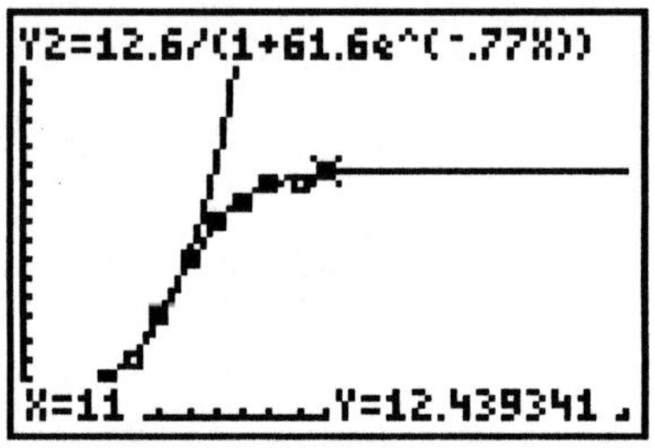

7. Type **20** and press [ENTER] to show the number of calcumite pairs (*y* value) after 20 generations of population growth.

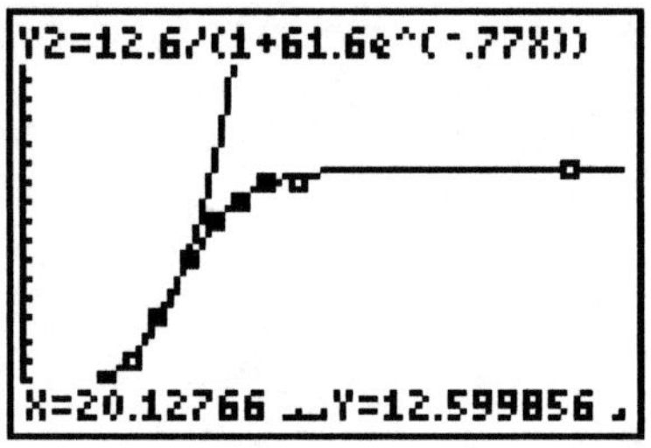

8. Record the approximate size of the population after 20 generations on the **Data Collection and Analysis** page.

9. To get an overview of how the two growth curves compare, reset the window of the calculator as shown.

   Do Not Change the ΔX Value.

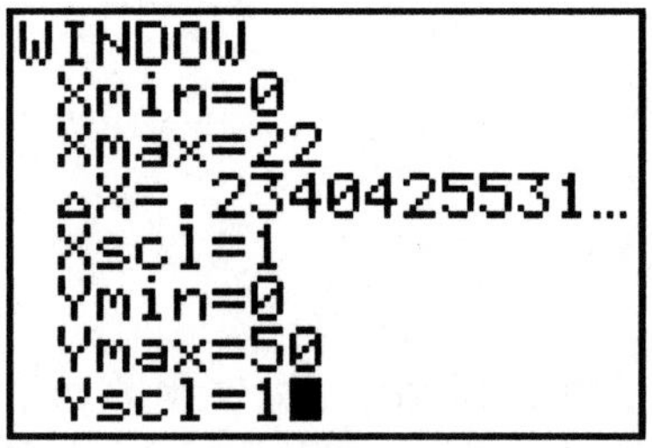

10. Press [GRAPH] to see the plot.

Answer Part II, Question 3 on the **Data Collection and Analysis** page.

Data Collection and Analysis

Name ______________________

Date ______________________

Activity 4: The Calcumites Are Coming!

Collecting the data

Generation	1	2	3	4	5	6	7	8	9	10
Pair Number	1	1	2	3	5					

Ideal Population Growth — Exponential

Generation	1	2	3	4	5	6	7	8	9	10
Pair Number	1	1	2	3	5	8	10	11	12	12

Limited Population Growth — Logistic

Ideal Population Growth

Slope (Interval: 2 – 3 Generations) = __________

Slope (Interval: 5 – 6 Generations) = __________

Slope (Interval: 8 – 9 Generations) = __________

Number of Calcumites after 20 Generations __________

Limited Population Growth

Slope (Interval: 2 – 3 Generations) = __________

Slope (Interval: 5 – 6 Generations) = __________

Slope (Interval: 8 – 9 Generations) = __________

Number of Calcumites after 20 Generations __________

Analyzing the data — Part I

1. What does the *slope* of a line on a population plot tell you?

____

____

2. Using the *slope* data, examine and describe the 2 – 3, 5 – 6, and 8 – 9 generation time intervals for the plot of *ideal* growth.

____

____

Analyzing the data — Part II

1. Describe the same generation time intervals for the plot of *limited* growth.

____

____

2. Which do you think is a more realistic model in nature? Why?

____

____

3. Which model - exponential or logistic - would be the better model to describe the following situations? Explain your answer in each case.

a. The velocity (speed) of a car going down an endless hill. (Graph velocity versus time.)

____

____

b. A virus spreading in your classroom. (Graph the number of people infected versus time.)

____

____

c. The number of TI-73 calculators sold. (Graph number sold versus time.)

____

____

Extension

Leonardo Fibonacci was an Italian mathematician who lived approximately 1175 – 1250. He is famous for discovering an interesting relationship between sequences of numbers that is now known as the *Fibonacci sequence*. The sequence starts with the following series of numbers: 0, 1, 1, 2, 3, 5, 8, 13, 21, 34, 55, 89, and so on. Starting with the third number, each number is equal to the sum of the two numbers preceding it. To study this interesting mathematical phenomenon, divide each number in the sequence by the number before it.

Use your knowledge of the natural world to find other species that contain examples of the Fibonacci sequence. Visualize the sequence by plotting your data. Notice that the data tends to stabilize around a certain number.

Teacher Notes

Activity 4

The Calcumites Are Coming!

Objectives

- To model the growth of a population
- To compare ideal population growth with a population whose growth is limited
- To use technology to find an exponential and logistic regression equation
- To use technology to plot an exponential and logistic model

Materials

- TI-73 graphing device
- Calcumite cut outs, pennies, or candy

Preparation

- The calcumite population examined in this activity is similar to the way in which Fibonacci examined a hypothetical rabbit population.
- Activity 8 will further examine Fibonacci sequences as applied to the spiral of a snail shell.
- While the exponential regression used to generate the plot of ideal population growth should be discussed, the logistic equation is beyond the scope of an introductory Algebra class. Yet, the general nature of the plot, with its changing slopes, can be studied to give the students an idea of how a population in nature often grows and stabilizes around the carrying capacity.
- It may be easier for some students to use manipulatives to better visualize the reproductive patterns of the calcumites. Here are a few suggestions:
  - Make copies of the calcumites on the final page of this activity.
  - Use coins or candy to simulate the calcumite populations.

Answers to Data Collection and Analysis questions

Collecting the data

Sample data:

Generation	1	2	3	4	5	6	7	8	9	10
Pair Number	1	1	2	3	5	*8*	*13*	*21*	*34*	*55*

Ideal Population Growth — Exponential

Generation	1	2	3	4	5	6	7	8	9	10
Pair Number	1	1	2	3	5	8	10	11	12	12

Limited Population Growth — Logistic

Ideal Population Growth

Slope (Interval: 2 – 3 Generations) = 1

Slope (Interval: 5 – 6 Generations) = 3

Slope (Interval: 8 – 9 Generations) = 13

Number of Calcumites after 20 Generations 5780

Limited Population Growth

Slope (Interval: 2 – 3 Generations) = 1

Slope (Interval: 5 – 6 Generations) = 3

Slope (Interval: 8 – 9 Generations) = 1

Number of Calcumites after 20 Generations 12

Analyzing the data — Part I

1. What does the *slope* of a line on a population plot tell you?

   The slope of a line on a population plot describes the rate of population growth.

2. Using the *slope* data, examine and describe the 2 – 3, 5 – 6, and 8 – 9 generation time intervals for the plot of *ideal* growth.

   For ideal growth, the rate of growth increases. Observe that the slope increases from 1 to 3 to 13 for the ranges examined.

Analyzing the data — Part II

1. Describe the same generation time intervals for the plot of *limited* growth.

   For limited growth, the rate of growth starts slow, speeds up, and then slows down. Observe that the slopes are 1, 3, and 1 for the ranges examined.

2. Which do you think is a more realistic model in nature? Why?

   The logistic model is more realistic because a population cannot increase in size, at a faster rate, forever. Limiting factors such as food availability, climate, disease, predation, and competition will eventually limit the size of any population.

3. Which model, exponential or logistic, would be the better model to describe the following situations? Explain your answer in each case.

a. The velocity (speed) of a car going down an endless hill. (Graph velocity versus time.)

Exponential: It will continuously accelerate down this hypothetically endless hill, assuming no air resistance.

b. A virus spreading in your classroom. (Graph the number of people infected versus time.)

Logistic: It will spread slowly at first as there are only a few infected individuals to pass it on. As more people are infected, they will quickly pass it on. As the number of uninfected individuals lessens, the likelihood of an infected individual contacting an uninfected individual decreases, and so the rate of spread slows down.

c. The number of TI-73 calculators sold. (Graph the number sold versus time.)

Logistic: The number of people purchasing the TI-73 calculators starts out slowly, increases as more people spread the good news, and slows down as few people are left who have not yet purchased them.

CALCUMITE CUT OUTS

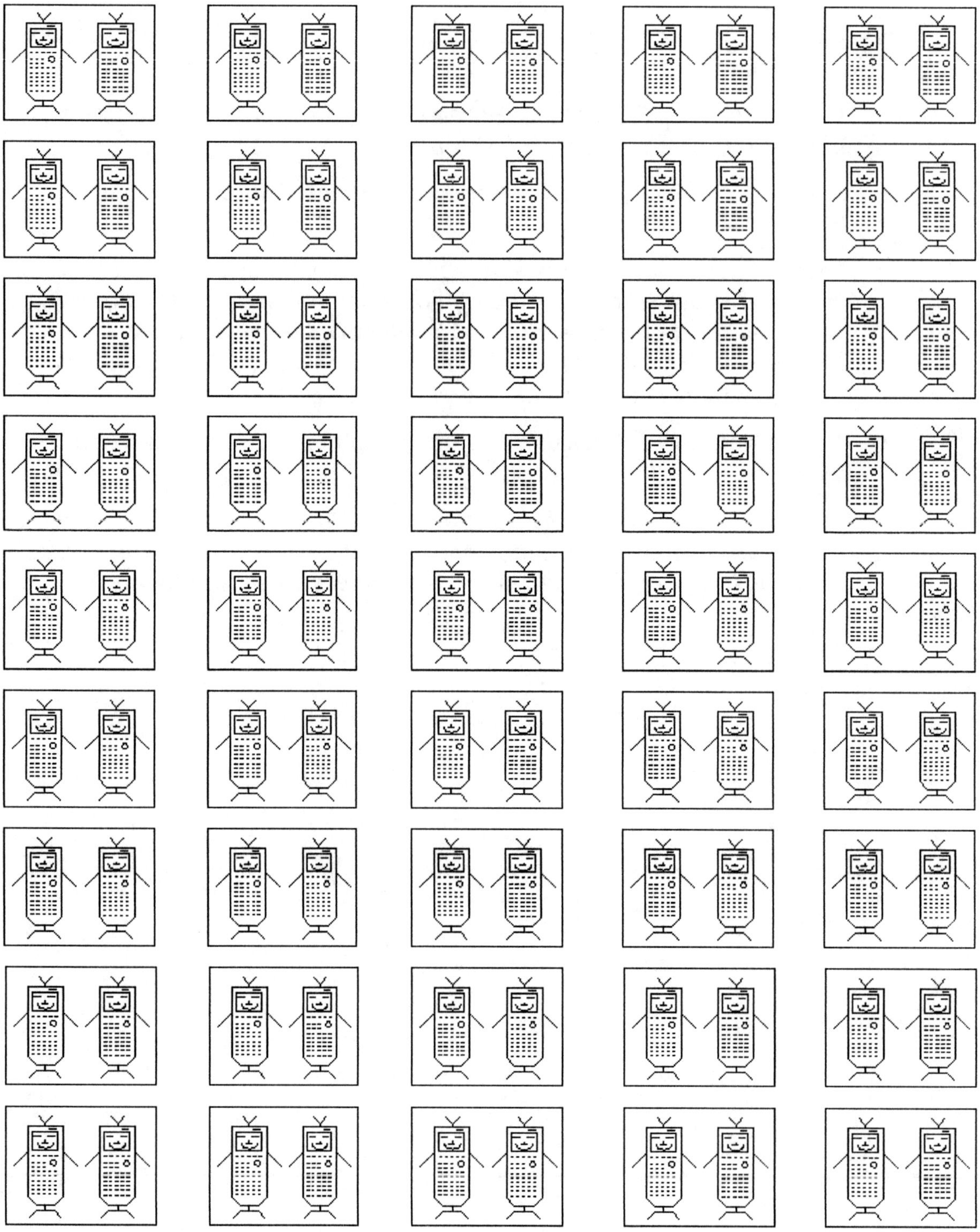

EXPLORATIONS

Activity 5

Give Me a Hand or Leaf Me Alone

Objectives

- To find the surface area of an irregularly shaped object by relating area to mass
- To find the *y* value of a function, given the *x* value
- To use technology to find a best fit line
- To use technology to plot a set of ordered pairs

Materials

- TI-73 graphing device
- Card stock paper (poster board, manila folders, or any heavy weight paper can be used)
- Scissors, one pair per student
- Scale or balance that measures in grams
- Ruler that measures in centimeters or inches, one per student
- Leaves of various sizes (at least one leaf per student)

Introduction

The idea of surface area is one of the most important concepts to understand in the biomedical sciences. Consider these examples. When you breathe, you must be able to absorb enough oxygen into your blood. Your highly compartmentalized lungs provide 70 square meters of surface area for oxygen absorption. That is about the size of the floor in your classroom. The surface area of the lining of your small intestines is 300 square meters, which is about the size of a tennis court. That allows you to efficiently absorb the nutrients from the food that you digest.

Surface area adaptations are found throughout the living world. Root hairs provide a tremendous surface area for water and mineral absorption, and the large surface area of leaves allows them to efficiently absorb sunlight.

Measuring the surface area of these irregularly shaped objects provides quite a challenge, one that is important enough to mathematically overcome.

Problem

There are formulas for finding the surface area of geometric figures such as a square, a rectangle, a triangle, or a circle. However, there are no such formulas for finding the surface area of an irregularly shaped object such as a hand or a leaf. How can you find the surface area of your hand or a leaf?

Collecting the data

1. Your teacher will assign you a specific length, between 1 and 18 centimeters, to use as the length for the side of a square. Cut a square of this side length from a piece of card stock paper.
2. Calculate the area of your square using the formula: $A = s^2$. Record the length of each side and the area of your square. Record the area on the square.
3. Use the scale to find the mass of your square in grams. Record the mass on the square.
4. Record the area and mass of your square on the **Data Collection and Analysis** page. Record all of the data for the class in the table on the **Data Collection and Analysis** page.

Setting up the TI-73

Before starting your data collection, make sure that the TI-73 has the STAT PLOTS turned OFF, Y= functions turned OFF or cleared, the MODE and FORMAT set to their defaults, and the lists cleared. See the Appendix for a detailed description of the general setup steps.

Entering the data in the TI-73

1. Press [LIST].

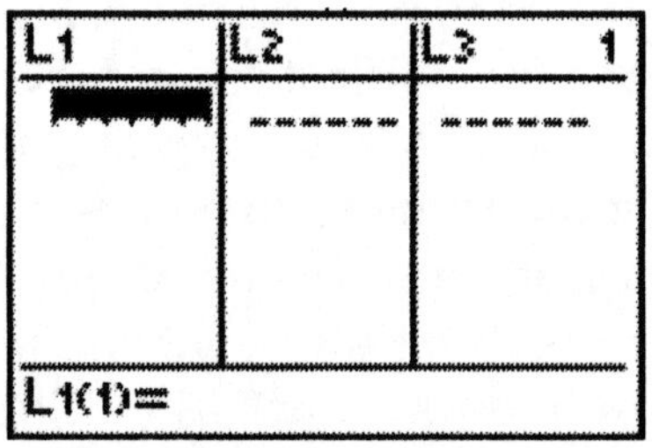

2. Enter the area of each square in **L1**.
3. Enter the mass of each square in **L2**.

   Note: Be sure to enter zero in both **L1** *and* **L2** *as your first entries.*

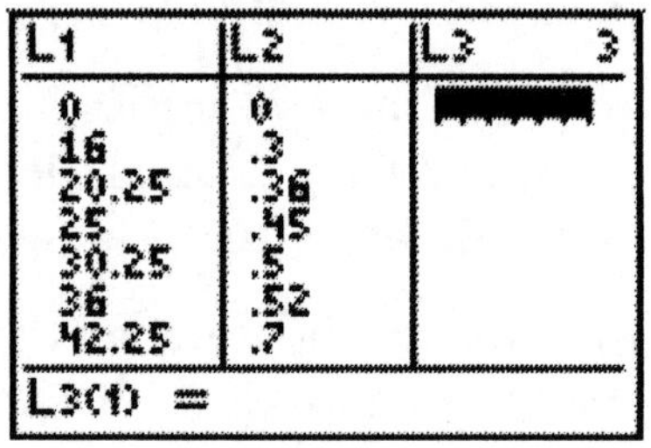

Setting up the window

1. Press [WINDOW] to set up the proper scale for the axes.
2. Set the **Xmin** value by identifying the minimum value in **L1**. Choose a number that is less than the minimum.

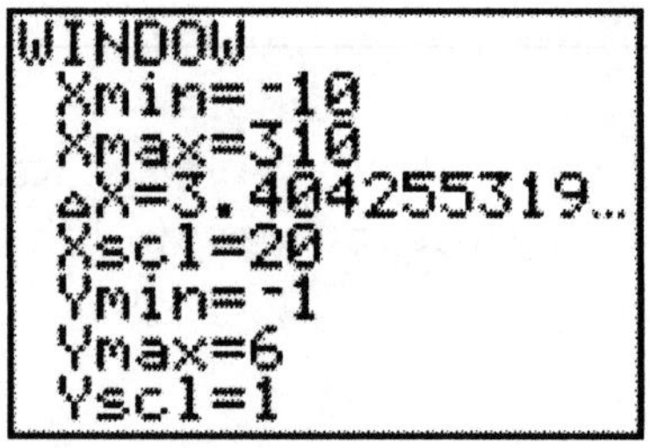

3. Set the **Xmax** value by identifying the maximum value in each list. Choose a number that is greater than the maximum. **Do Not Change the ΔX Value.** Set the **Xscl** to **20**.

4. Set the **Ymin** value by identifying the minimum value in **L2**. Choose a number that is less than the minimum.

5. Set the **Ymax** value by identifying the maximum value in **L2**. Choose a number that is greater than the maximum. Set the **Yscl** to **1**.

Graphing the data: Setting up a scatter plot

1. Press [2nd] [PLOT]. Select **1:Plot1** by pressing **1** or [ENTER].

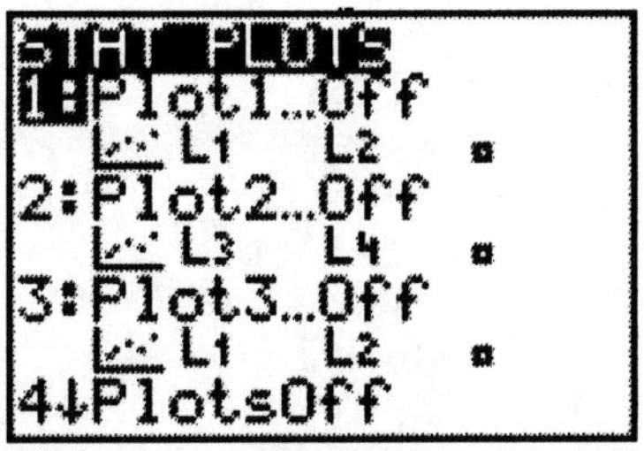

2. Set up the plot as shown by pressing [ENTER] [▼] [ENTER] [▼] [2nd] [STAT] **1:L1** [▼] [2nd] [STAT] **2:L2** [▼] [ENTER].

3. Press [TRACE]. Use [◄] and [►] to move between the points.

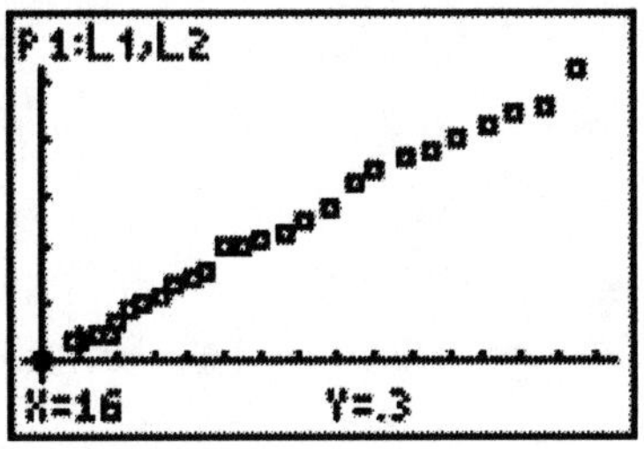

Analyzing the data

Finding a best fit line

1. Find a linear regression equation for the data. Press [2nd] [STAT] [◄] to move the cursor to the **CALC** menu.

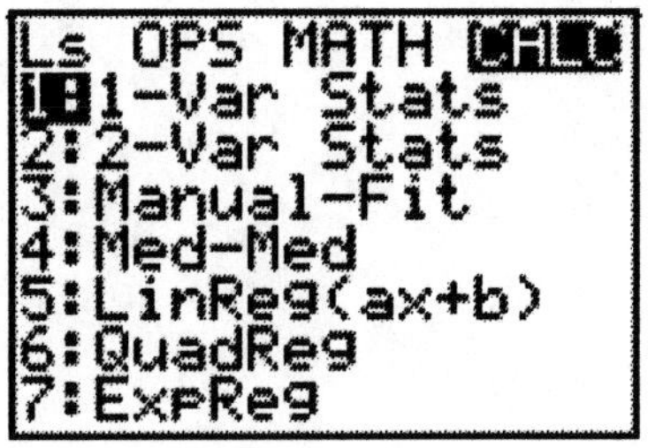

2. Select **5:LinReg(ax + b)** by pressing **5.**

```
LinReg(ax+b)
```

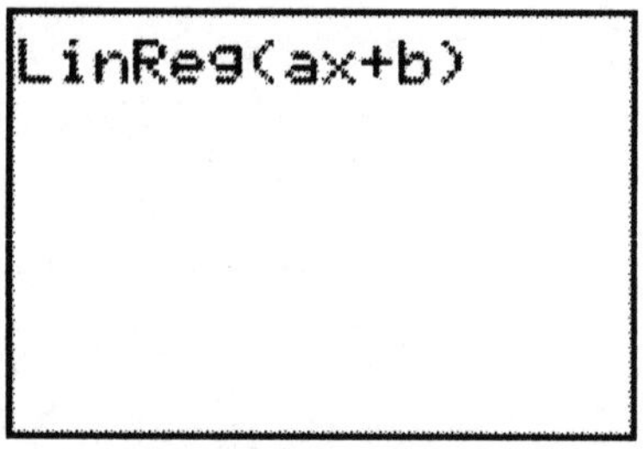

3. Press [2nd] [STAT] **1:L1** [,] [2nd] [STAT] **2:L2** [,].

```
LinReg(ax+b) L₁,
L₂,
```

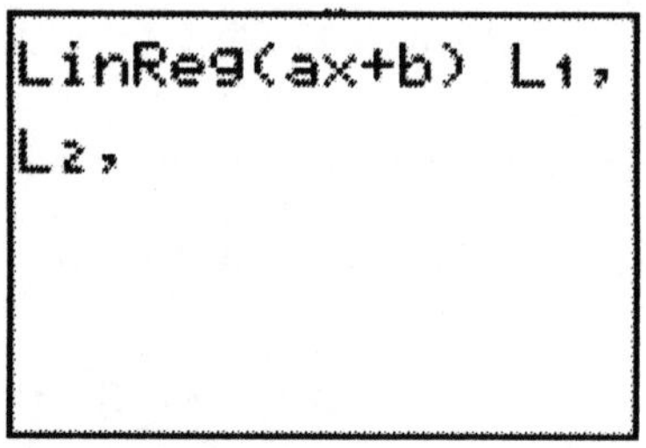

4. Press [2nd] [VARS]. Select **2:Y-Vars** by pressing **2.**

```
FUNCTION
1:Y1
2:Y2
3:Y3
4:Y4
5:FnOn
6:FnOff
```

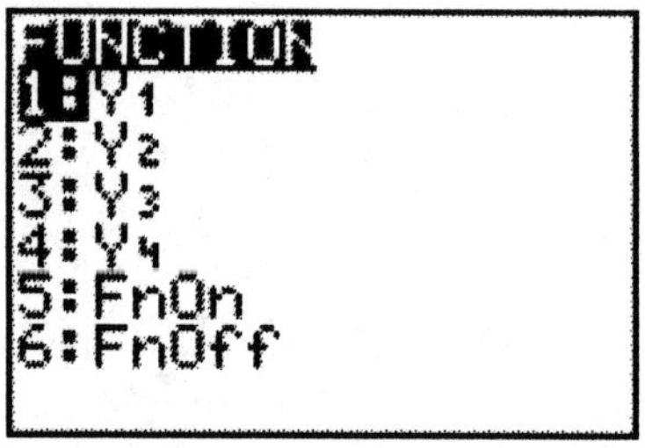

5. Select **1:Y1** by pressing **1** or [ENTER].

```
LinReg(ax+b) L₁,
L₂,Y₁
```

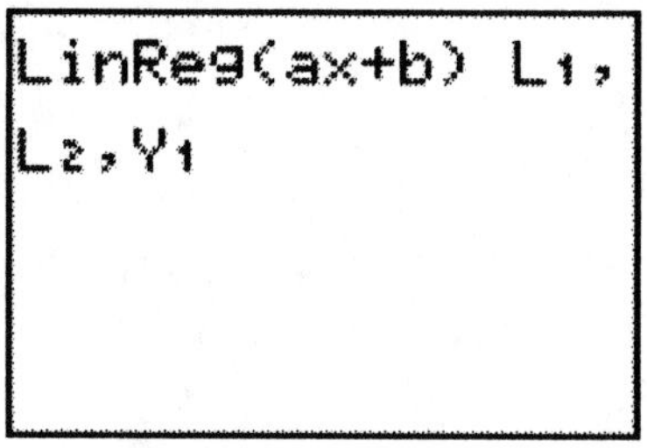

6. Press [ENTER] to calculate the linear regression. The function is pasted in **Y1.**

```
LinReg
y=ax+b
a=.0175469115
b=.0400005813
```

7. Press [Y=] to see the function.

```
Plot1 Plot2 Plot3
\Y1■.0175469148
994X+.0400005812
868
\Y2=
\Y3=
```

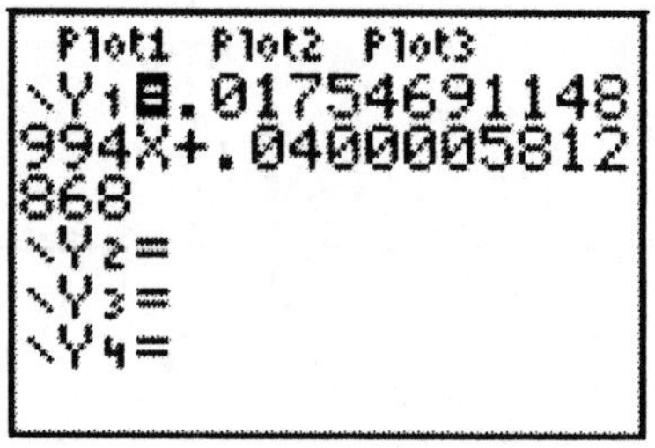

8. Press [GRAPH] to see the graph of the best fit line.

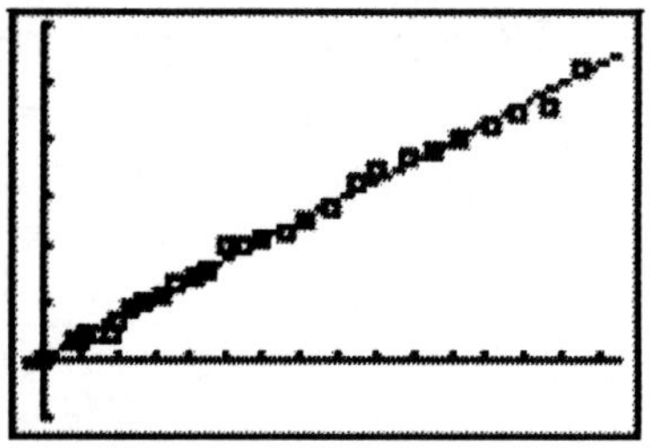

Answer questions 1 through 4 on the **Data Collection and Analysis** page.

Finding the area of your hand and a leaf

1. Trace your hand and a leaf on a sheet of card stock paper. Cut out each of the tracings. Find the mass in grams of your cut out hand and your cut out leaf.

2. Find the surface area of your hand. Press [Y=] and [▼] until the cursor is on **Y2**. Enter the mass of your cut out hand in **Y2**.

3. Turn **Plot1** off. Press [▲] until the cursor is on **Plot1**. Press [ENTER] to turn it off. Press [▼] and notice that **Plot1** is no longer highlighted. Press [GRAPH] to see the intersection of the two lines.

   Note: *You may have to change the window to see the intersection of the lines.*

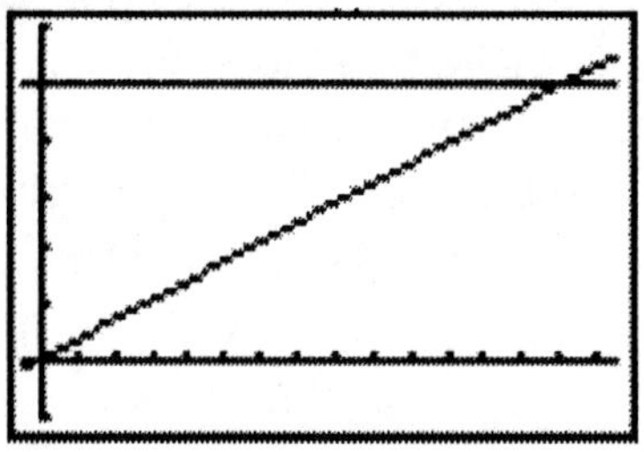

   Find the coordinates of the point of intersection of the two lines. Draw a vertical line that passes through the point of intersection.

4. Press [DRAW].

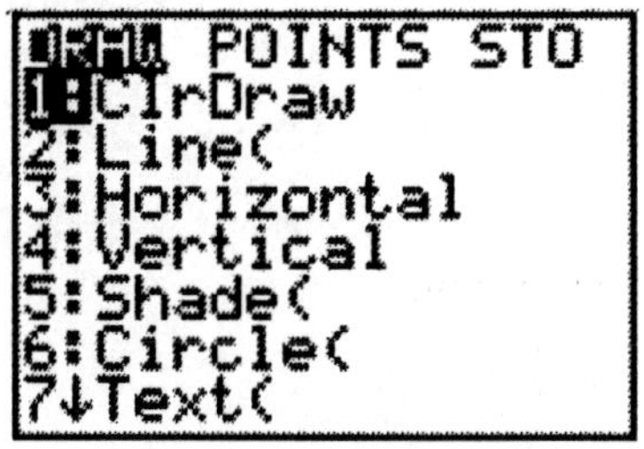

5. Select **4:Vertical** by pressing **4**.

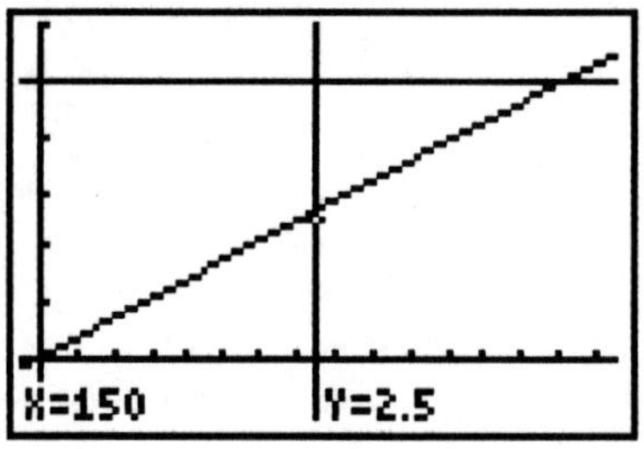

6. Press [▶] to move the vertical line until you reach the point of intersection.

   Note: *The x value is the approximate area.*

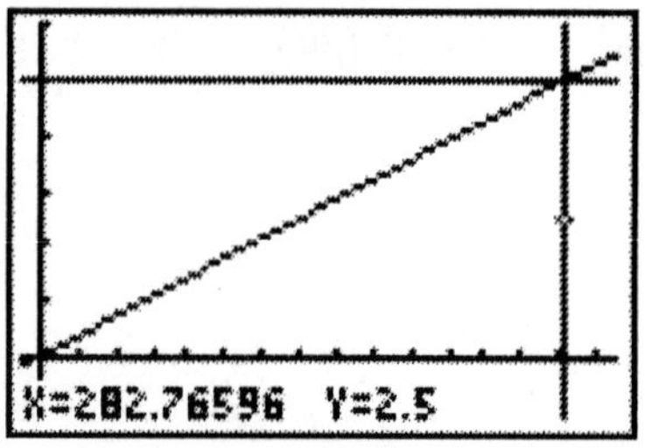

7. Use the **TABLE** to confirm the area. Press [2nd] [TBLSET] and set up the **TABLE** like the one shown to the right. Use a value close to the *x* value that you found above as your **TblStart**.

8. Press [2nd] [TABLE] to view the **TABLE**.

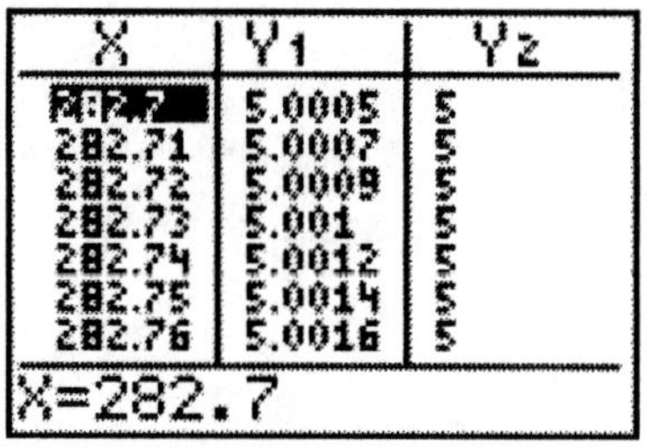

X
Y1
Y2
282.71
282.72
282.73
282.74
282.75
282.76
5.0005
5.0007
5.0009
5.001
5.0012
5.0014
5.0016
5
5
5
5
5
5
5
X=282.7

9. If necessary, use [▲] and [▼] to scroll the table to find the x value that corresponds to the mass of your cut out hand. Record the point of intersection found in Step 5 on the **Data Collection and Analysis** page.

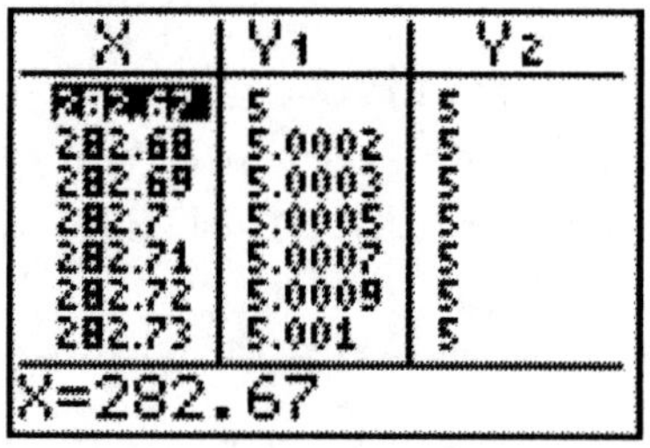

X
Y1
Y2
282.68
282.69
282.7
282.71
282.72
282.73
5
5.0002
5.0003
5.0005
5.0007
5.0009
5.001
5
5
5
5
5
5
5
X=282.67

Answer questions 6 through 9 on the **Data Collection and Analysis** page.

Follow steps 2 through 9 above to find the surface area of the leaf. Answer questions 10 through 13 on the **Data Collection and Analysis** page.

Data Collection and Analysis

Name ______________________

Date ______________________

Activity 5: Give Me a Hand or Leaf Me Alone

Collecting the data

Record your data in the table below.

Student No.	Length of side of square (cm)	Area of square (cm^2)	Mass of square (grams)
1	0		
2	4		
3	4.5		
4	5		
5	5.5		
6	6		
7	6.5		
8	7		
9	7.5		
10	8		
11	8.5		
12	9		
13	9.5		
14	10		
15	10.5		
16	11		
17	11.5		
18	12		
19	12.5		
20	13		
21	13.5		
22	14		
23	14.5		
24	15		

Analyzing the data

1. The *slope* of the linear regression line is ______________________.
2. Explain what the *slope* represents.

____

____

3. The *y*-intercept of the line is ______________________.
4. Explain what the *y*-intercept represents.

____

____

5. Record the coordinates of the point of intersection of your two lines for the hand data.

____

6. What does the *x* value represent?

____

____

7. What does the *y* value represent?

____

____

8. To find the approximate surface area of your hand, double the value that represents the surface area in number 7. You are doubling the surface area of your hand to approximate adding the top and bottom (neglecting the sides) of your hand. The total surface area of your hand is:

____

9. What is the surface area of both of your hands?

____

10. Record the coordinates of the point of intersection of your two lines for the leaf data.

____

11. What does the *x* value represent?

____

12. What does the *y* value represent?

____

13. To find the approximate surface area of the leaf, double the value that represents the surface area. The total surface area of the leaf is:

____

Extension

Doctors sometimes use body surface area to determine the dosage of medicine to prescribe to their patients. There are several formulas for calculating the Body Surface Area (BSA). Boyd and Mosteller developed this formula:

$$\text{BSA} = \frac{\sqrt{\text{Height (cm)} \bullet \text{Weight (kg)}}}{3600}$$

Use the formula to calculate your body surface area.

Teacher Notes

Activity 5

Give Me a Hand or Leaf Me Alone

Objectives

- To find the surface area of an irregularly shaped object by relating area to mass
- To find the y value of a function, given the x value
- To use technology to find a best fit line
- To use technology to plot a set of ordered pairs

Materials

- TI-73 graphing device
- Card stock paper (poster board, manila folders, or any heavy weight paper can be used)
- Scissors, one pair per student
- Scale or balance that measures in grams
- Ruler that measures in centimeters or inches, one per student
- Leaves of various sizes (at least one leaf per student)

Preparation

- The paper used for cutting the hand and leaf tracings can be tag board, chart paper, folders, or any heavy paper.
- Collect all of the squares with the area and mass recorded on them. Before proceeding to the **Finding the area of your hand and a leaf** section, use the squares to estimate the area of the cut out hand and leaves. Allow the students to place their cut out hand or leaf on the square that is closest in area to their cut out. Estimate the area of their cut out hand or leaf.
- Make sure you adjust the window when finding the intersection of the two lines.

Answers to Data Collection and Analysis questions

Collecting the data

Sample data for a leaf:

Area of square (cm^2)	Mass of square (grams)
0	0
16	.3
20.25	.36
25	.45
30.25	.5
36	.52
42.25	.7
49	.97
56.25	1
64	1.2
72.25	1.33
81	1.5
90.25	1.56
100	2.05
110.25	2.1
121	2.18
132.25	2.25
144	2.53
156.25	2.78
169	3.2
182.25	3.42
196	3.61
210.25	3.75
225	3.95
240.25	4.17
256	4.41
272.25	4.55
289	5.2

Analyzing the data

1. The *slope* of the linear regression line is ______________________________ .

   Answers may vary.
   The slope of the linear regression line is 0.0175.

2. Explain what the *slope* represents.

   The slope represents the number of grams per square centimeter of area. For the data presented, for every square centimeter increase in area the mass increases by about 0.0175g.

3. The *y*-intercept of the line is ______________________________ .

   The y-intercept is 0.0400.

4. Explain what the *y*-intercept represents.

   The y-intercept indicates that a cut out with an area of zero has a mass of 0.0400g. Of course, this is not the case. Point out to students that this is a model and the y-intercept is close to zero.

5. Record the coordinates of the point of intersection of your two lines for the hand data.

   Answers may vary.

6. What does the *x* value represent?

   The coordinate x represents the area of the cut out hand.

7. What does the *y* value represent?

   The coordinate y represents the mass of the cut out hand.

8. To find the approximate surface area of your hand, double the value that represents the surface area in number 7. You are doubling the surface area of your hand to approximate adding the top and bottom (neglecting the sides) of your hand. The total surface area of your hand is: ______________ .

   Answers may vary.

9. What is the surface area of both of your hands?

   Answers may vary.

10. Record the coordinates of the point of intersection of the two lines for the leaf data.

    Answers may vary.

11. What does the *x* value represent?

    The coordinate x represents the area of the cut out leaf.

12. What does the *y* value represent?

The coordinate y represents the mass of the cut out leaf.

13. To find the approximate surface area of the leaf, double the value that represents the surface area. The total surface area of the leaf is: __________.

Answers may vary.

EXPLORATIONS

Activity 6

You're So Dense

Objectives

- To investigate the relationship between mass and volume
- To find the x value of a function, given the y value
- To find the y value of a function, given the x value
- To use technology to find a linear regression
- To use technology to plot a set of ordered pairs

Materials

- TI-73 graphing device
- 50 ml graduated cylinder
- Balance
- Pennies (25 pre-1982 and 25 post-1984)

Introduction

Have you ever picked up an object that you thought was heavy, and it was light? For example, if you saw a solid metal block that is 10 cm^3 you might expect it to be difficult to lift. If that block were aluminum, its mass would only be 2.7 kg (or a little under 6 pounds). On the other hand, if the metal block were platinum, its mass would be 21.5 kg (or 47 pounds). The reason for this difference is that metals have different densities. The density of an object is a measure of the mass divided by the volume.

Problem

How can you use the density of pennies to predict the dates they were made?

Collecting the data

1. Your teacher will give you pennies. Separate the pennies into two piles: those that were made before 1982 and those that were made after 1984. For steps 2 through 8, use the pennies that were made before 1982. Repeat steps 2 through 8 using the pennies that were made after 1984.

2. Fill a graduated cylinder with water to the 20 ml mark. Using a balance, determine the mass of the graduated cylinder with the water.

3. Add five pennies to the graduated cylinder. Determine the difference in the volume by subtracting the original 20 ml from the new volume. This difference is the volume of the pennies that you have added. Record the volume of the five pennies on the **Data Collection and Analysis** page.

   Note: *It is assumed that 0 pennies has a volume of 0 ml and a mass of 0 grams. These values have already been recorded on the* **Data Collection and Analysis** *page.*

4. Determine the difference in the mass by subtracting the original mass from the new mass. The difference is the mass of the pennies that you have added. Record the mass of the five pennies on the **Data Collection and Analysis** page.

5. Add a second set of five pennies (for a total of 10) to the graduated cylinder. Follow the directions in steps 3 and 4 to determine the *total* volume and *total* mass of the 10 pennies. Record the volume and mass of the 10 pennies on the **Data Collection and Analysis** page.

6. Add another five pennies (for a total of 15) to the graduated cylinder. Follow the directions in steps 3 and 4 to determine the *total* volume and *total* mass of the 15 pennies. Record the volume and mass of the 15 pennies on the **Data Collection and Analysis** page.

7. Add another five pennies (for a total of 20) to the graduated cylinder. Follow the directions in steps 3 and 4 to determine the *total* volume and *total* mass of the 20 pennies. Record the volume and mass of the 20 pennies on the **Data Collection and Analysis** page.

8. Add another 5 pennies (for a total of 25) to the graduated cylinder. Follow the directions in steps 3 and 4 to determine the *total* volume and *total* mass of the 25 pennies. Record the volume and mass of the 25 pennies on the **Data Collection and Analysis** page.

Setting up the TI-73

Before starting your data collection, make sure that the TI-73 has the STAT PLOTS turned OFF, Y= functions turned OFF or cleared, the MODE and FORMAT set to their defaults, and the lists cleared. See the Appendix for a detailed description of the general setup steps.

Entering the data in the TI-73

1. Press [LIST].

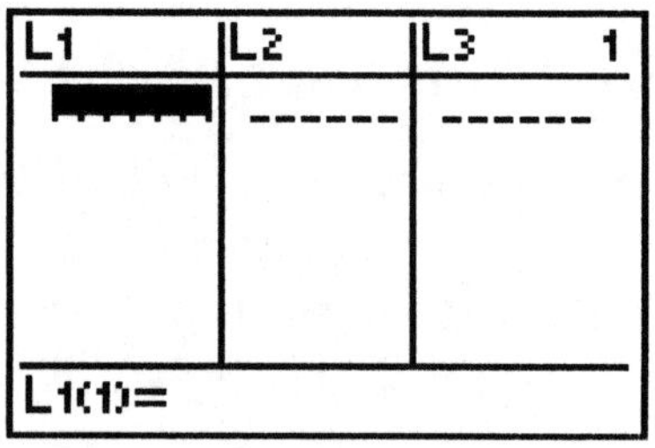

2. Enter the volume of the pre-1982 pennies in **L1**.

3. Enter the mass of the pre-1982 pennies in **L2**.

4. Enter the volume of the post-1984 pennies in **L3**.

5. Enter the mass of the post-1984 pennies in **L4**.

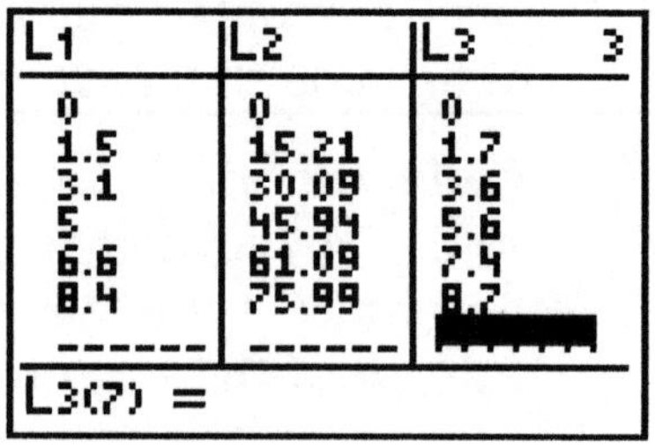

Setting up the window

1. Press [WINDOW] to set up the proper scale for the axes.
2. Set the **Xmin** value by identifying the minimum value in **L1**. Choose a number that is less than the minimum.
3. Set the **Xmax** value by identifying the maximum value in each list. Choose a number that is greater than the maximum. **Do Not Change the ΔX Value.** Set the **Xscl** to **2**.

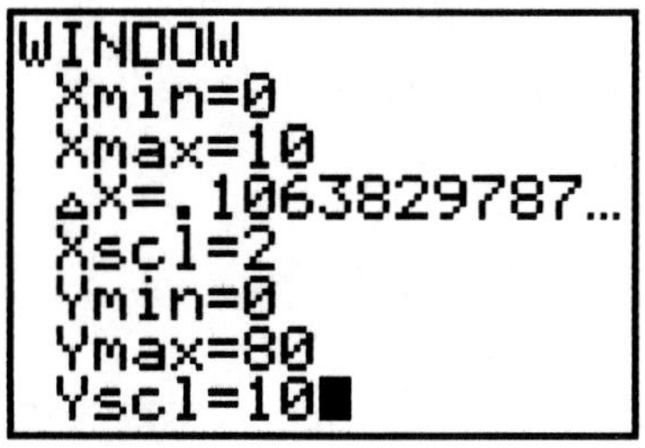

4. Set the **Ymin** value by identifying the minimum value in **L2**. Choose a number that is less than the minimum.
5. Set the **Ymax** value by identifying the maximum value in **L2**. Choose a number that is greater than the maximum. Set the **Yscl** to **10**.

Graphing the data: Setting up the scatter plots

To analyze the data, you will need to set up a scatter plot for each set of data and then model that data by graphing a line of best fit (linear regression). You will then use the data that you collected to compare the pre-1982 to the post-1984 pennies.

1. Press [2nd] [PLOT]. Select **1:Plot1** by pressing **1** or [ENTER].

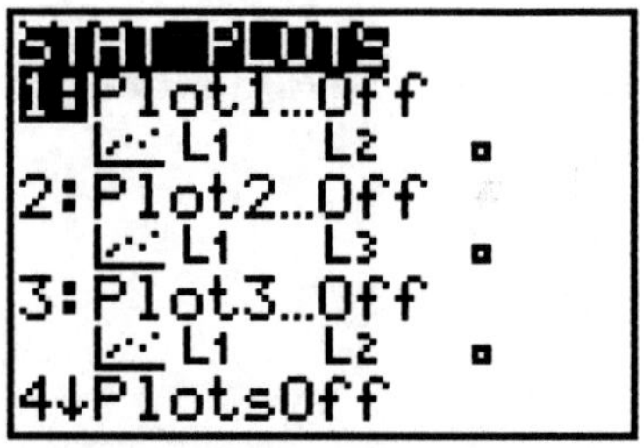

2. Set up the plot as shown by pressing [ENTER] [▼] [ENTER] [▼] [2nd] [STAT] **1:L1** [▼] [2nd] [STAT] **2:L2** [▼] [ENTER].

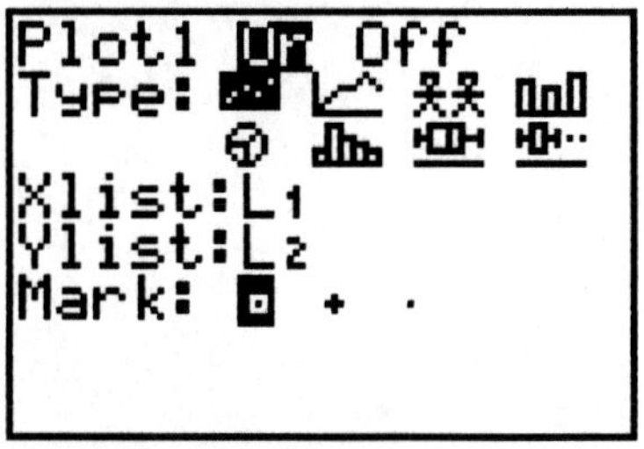

3. Press [2nd] [PLOT]. Select **2:Plot2** by pressing **2**.

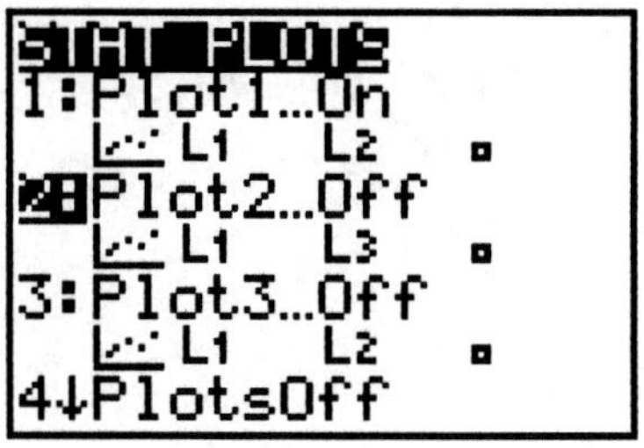

4. Set up the plot as shown by pressing [ENTER] [▼] [ENTER] [▼] [2nd] [STAT] **3:L3** [▼] [2nd] [STAT] **4:L4** [▼] [▶] [ENTER].

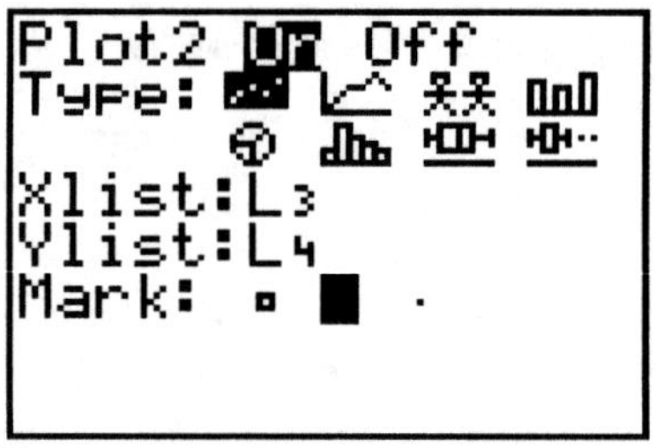

5. Press [GRAPH] to see the plots.

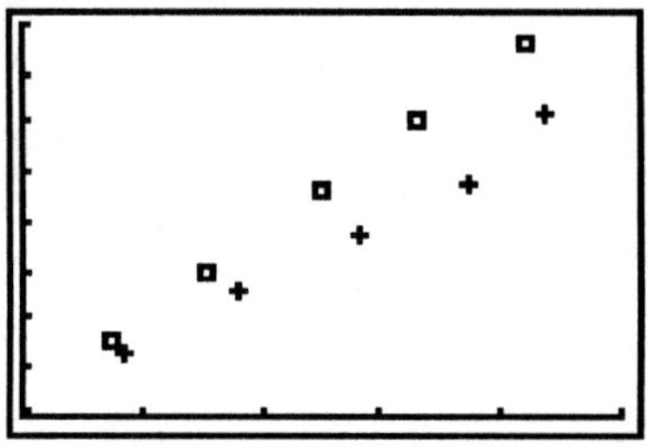

Analyzing the data

Finding a linear regression

1. Plot a linear regression for the pre-1982 penny data. Press [2nd] [STAT] [◀] to move the cursor to the **CALC** menu.

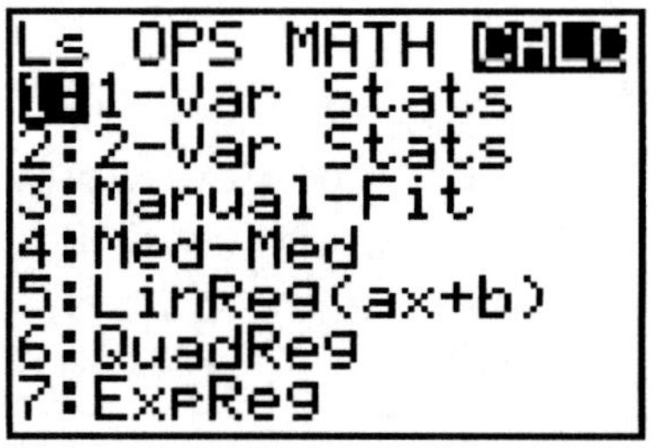

2. Select **5:LinReg(ax+b)** by pressing **5**.

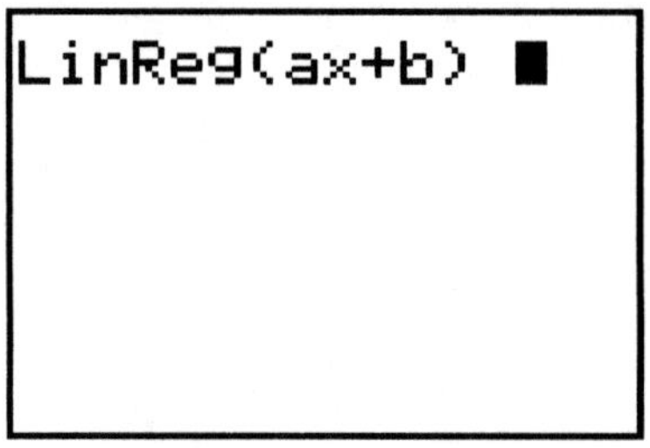

3. Press [2nd] [STAT] **1:L1** [,] [2nd] [STAT] **2:L2** [,].

4. Press [2nd] [VARS]. Select **2:Y-Vars** by pressing **2**.

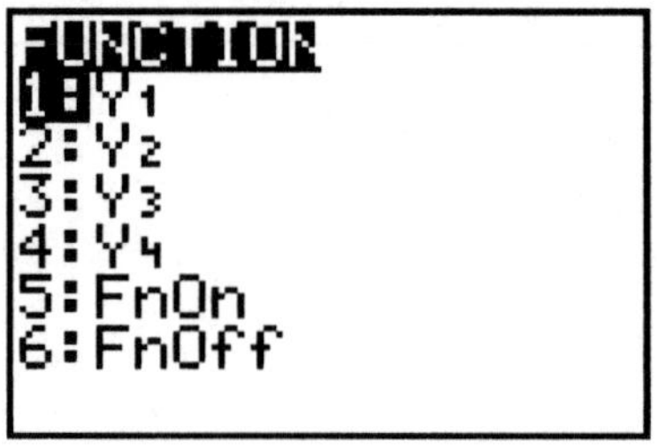

5. Select **1:Y1** by pressing **1** or [ENTER].

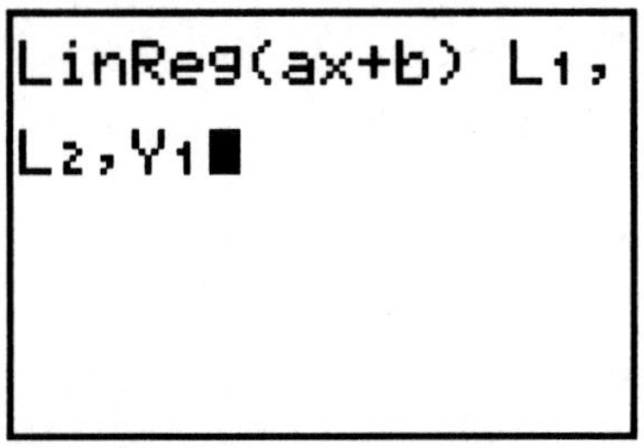

6. Press [ENTER] to calculate the linear regression. The function is pasted in **Y1**.

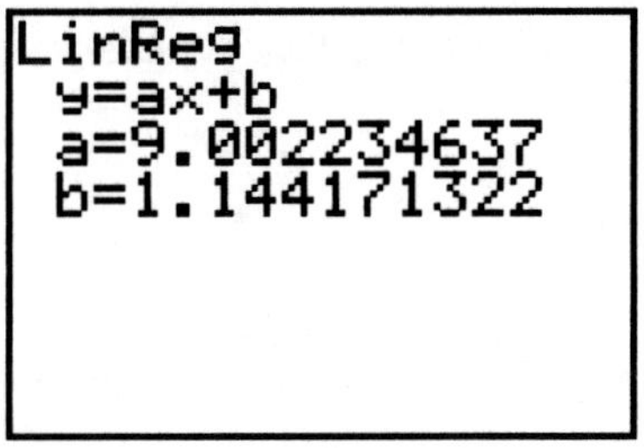

7. Press [GRAPH] to see the linear regression model.

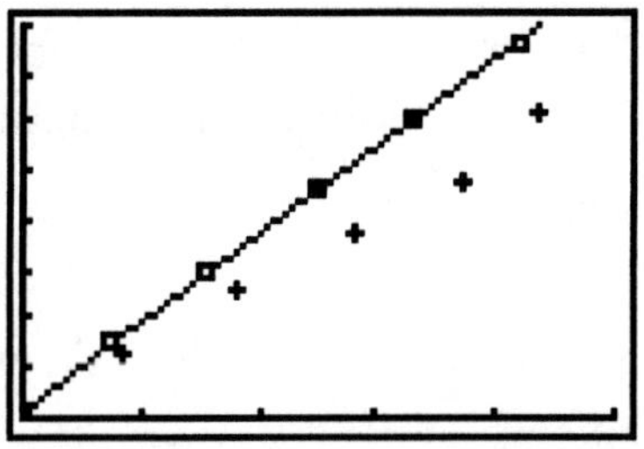

8. Plot a linear regression for the post-1984 penny data. Press [2nd] [STAT] [◄] to move the cursor to the **CALC** menu.

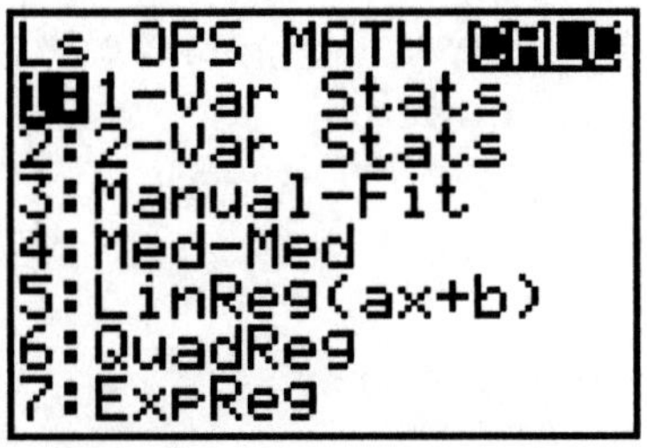

9. Select **5:LinReg(ax+b)** by pressing **5**.

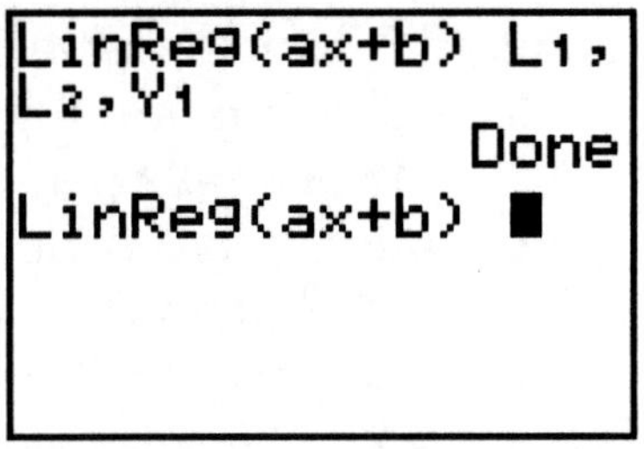

10. Press [2nd] [STAT] **3:L3** [,] [2nd] [STAT] **4:L4** [,].

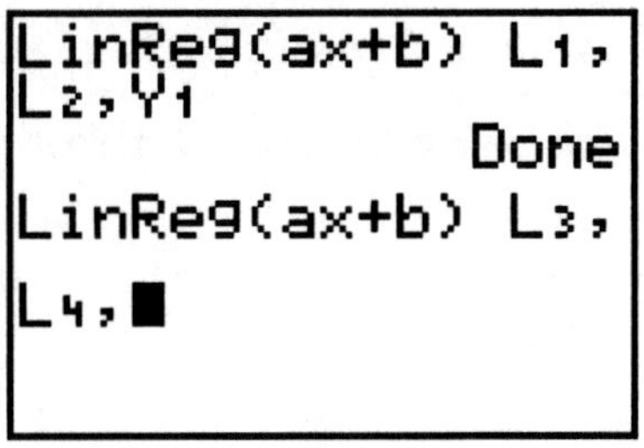

11. Press [2nd] [VARS]. Select **2:Y-Vars** by pressing **2**.

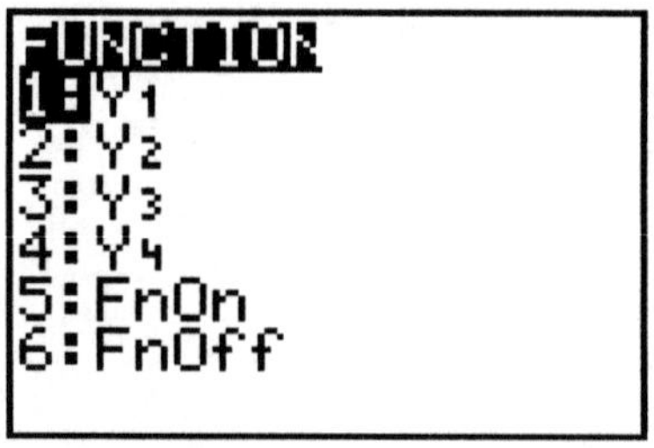

12. Select **2:**$\mathbf{Y_2}$ by pressing **2**.

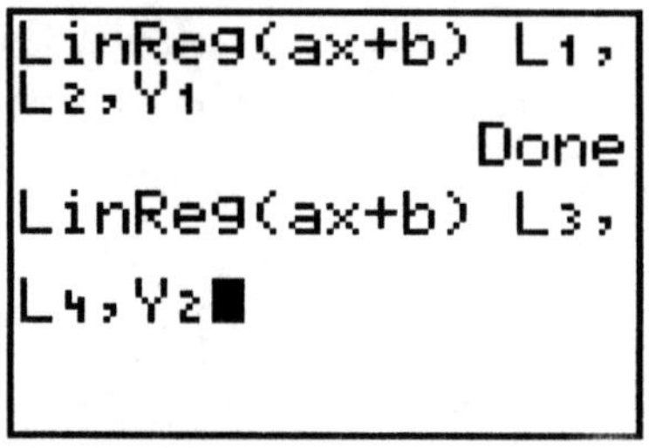

13. Press [ENTER] to calculate the linear regression. The function is pasted in $\mathbf{Y_2}$.

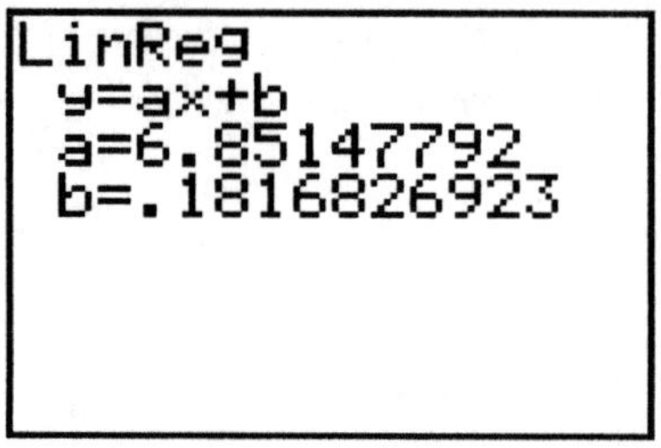

14. Press [GRAPH] to see the linear regression model.

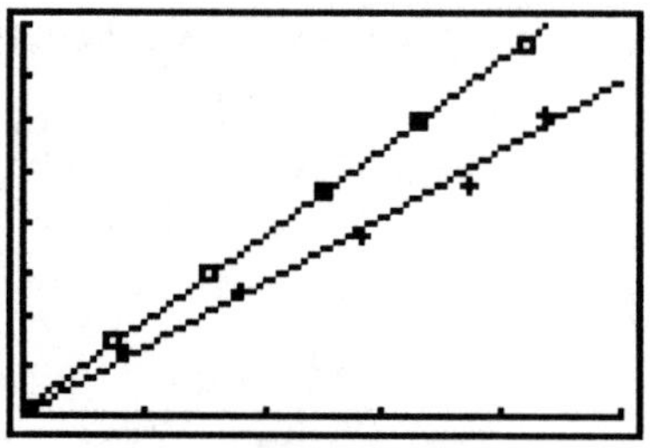

15. Press [Y=] and observe the two equations. $\mathbf{Y_1}$ is the equation that describes the data for the pre-1982 pennies and $\mathbf{Y_2}$ is the equation that describes the data for the post-1984 pennies. The equations are for linear regressions and are therefore in the $Y = AX + B$ format where A is the *slope* and B is the y-intercept.

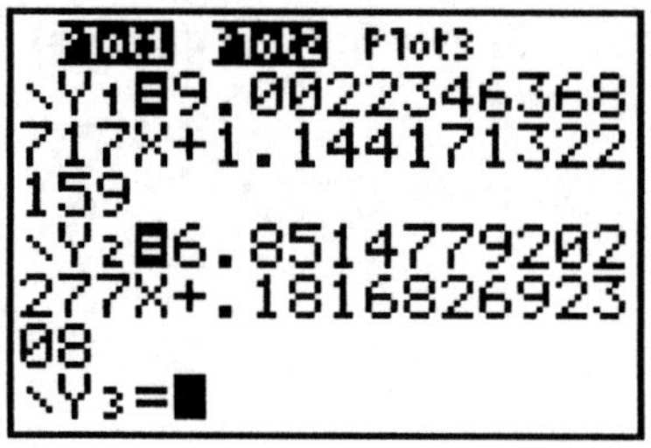

Answer questions 1 through 5 on the **Data Collection and Analysis** page.

Determining the density of pennies

You can determine the dates that the pennies were minted based on their density.

1. Your teacher will give you 15 pennies, but will not tell you whether they are pre-1982 or post-1984. They will all be from one of the two date categories (that is, either all will be pre-1982 or post-1984). Measure their mass and determine their volume by water displacement as you did earlier in the activity.
2. Determine the density by dividing the mass by the volume. Which category of penny do you have? Confirm your findings by checking the dates.

Predicting the volume of pennies

You can predict the volume of pennies if you know their mass.

1. Your teacher will give you 15 pennies and tell you whether they are pre-1982 or post-1984. Using the equation $D = \frac{M}{V}$ and the density determined earlier, measure their mass with a balance and mathematically determine their volume.
2. Verify your answer using the TI-73. Press [Y=] and move the cursor to **Y3** and enter the mass of the pennies.

   (In this example, assume that the mass is 36 kg and that the pennies are pre-1982. The teacher may give you a different number of pennies with a different mass.)

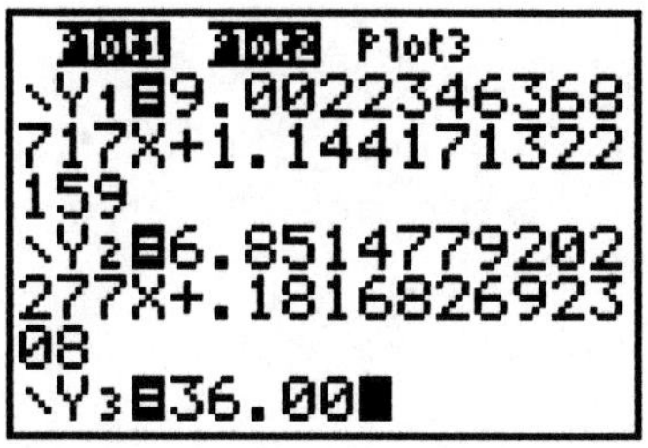

3. Press [GRAPH] to see the intersection of the lines.

   The x values of the points where the lines intersect relate to the volumes for 15 pennies.

   Use the **Table** function of the TI-73 to determine the coordinates of the points of intersection.

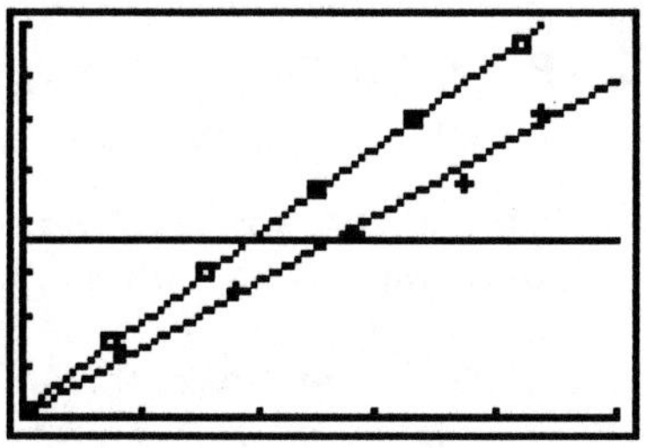

4. Press [2nd] [TBLSET]. Type **0** for **TblStart**. Press [▼] **1** to set the **ΔTbl** value.

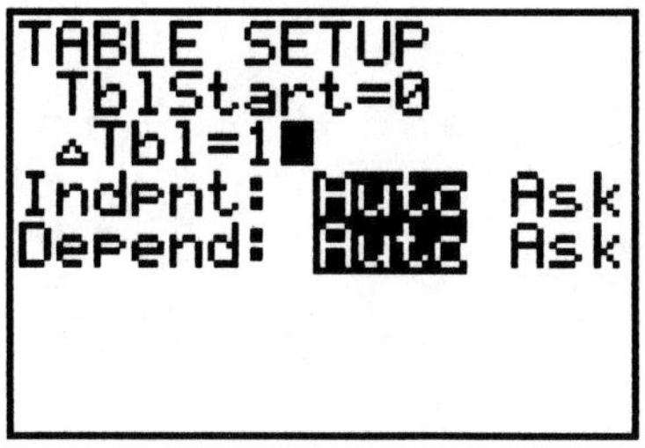

5. Press [2nd] [TABLE]. If necessary, use [▼] or [▲] to scroll the table.

    Note: *For this example, in the **Y1** column, 36 g falls between 28.151 and 37.153 which corresponds to 3 and 4 mL. Based on this information, the table will be readjusted.*

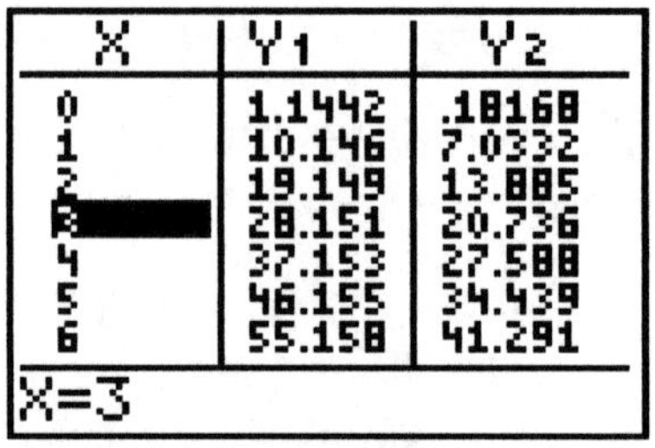

6. Press [2nd] [TBLSET]. Enter the results from Step 5 for the **TblStart** value. Press [▼] **0.1** to set the **ΔTbl** value.

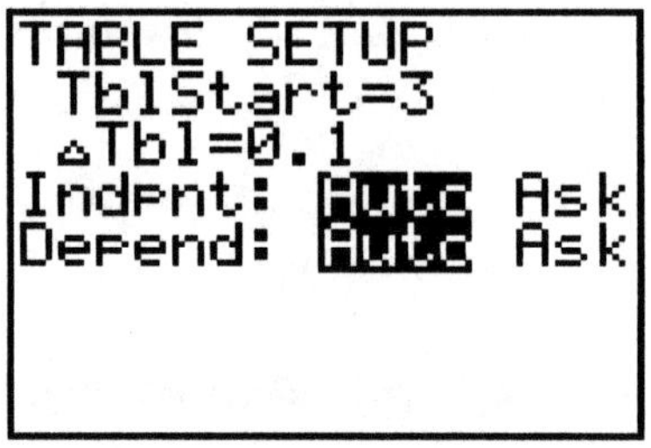

7. Press [2nd] [TABLE]. If necessary, use [▼] or [▲] to scroll the table.

    Note: *For this example, in the **Y1** column, 36 g falls between 35.353 and 36.253 which corresponds to 3.8 and 3.9 mL. Based on this information, the table will be readjusted.*

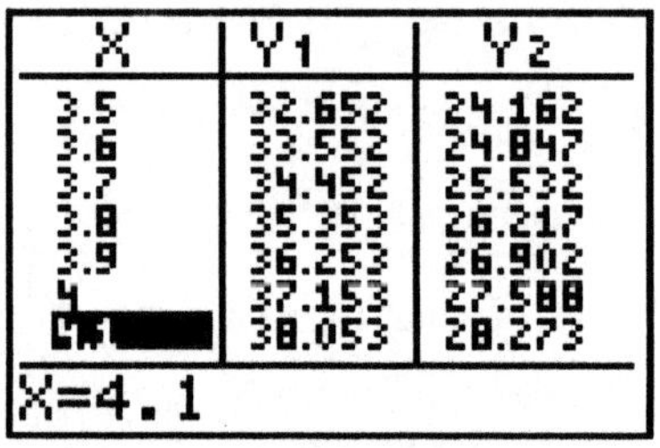

8. Press [2nd] [TBLSET]. Enter the results from Step 7 for the **TblStart** value. Press [▼] **0.01** to set the **ΔTbl** value.

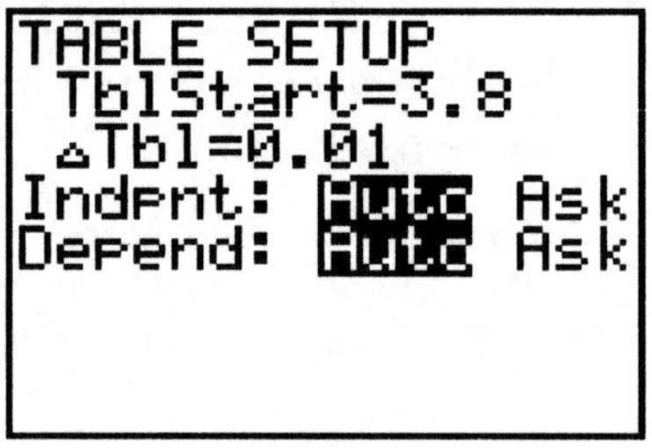

9. Press [2nd] [TABLE]. If necessary, use [▼] or [▲] to scroll the table.

    Note: *The data used to construct the linear model had volumes measured to the nearest tenth of a mL. Therefore, the volume of the pennies is reported to the same level of precision. From the table, 36 g falls between 35.983 and 36.073 which corresponds to 3.87 and 3.88 mL. Rounding to the nearest tenth of a mL, the intersection point is (3.9, 36).*

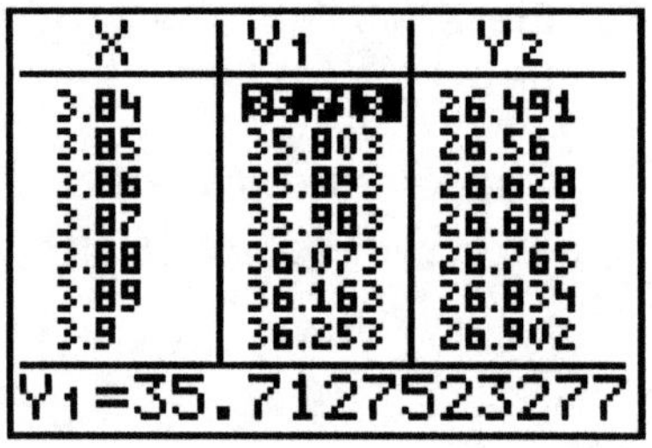

10. To verify this graphically, use the **DRAW** function. Press [DRAW]. Select **4:Vertical** by pressing **4**.

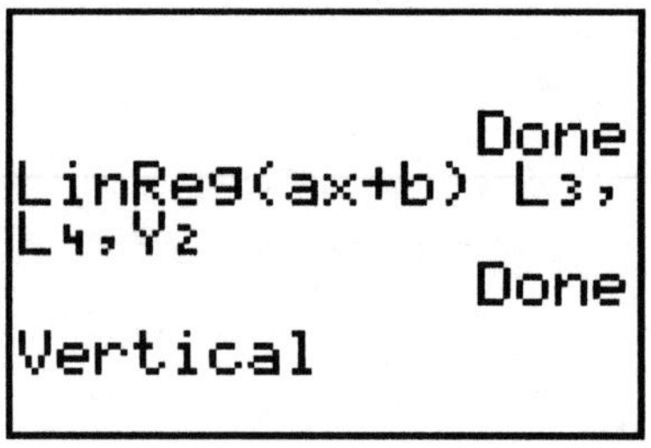

11. Type your results from Step 9 (in this example, **3.9**), and press [ENTER].

    Note: *The coordinates of the point on the linear model for pre-1982 pennies where all of the lines intersect is defined, for this example, by the vertical drawn at x=3.9 and the horizontal at y=36.*

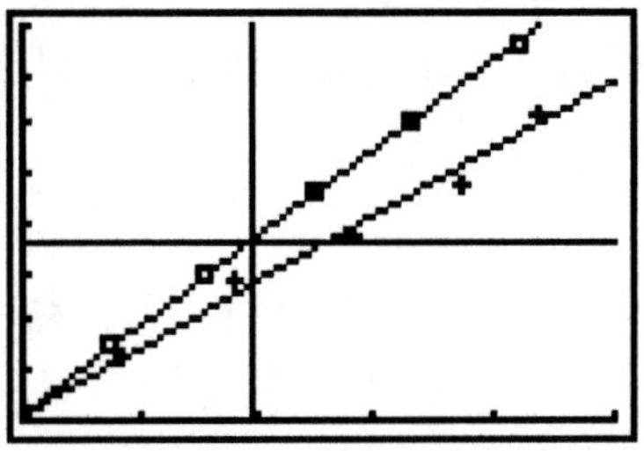

12. The coordinates of the intersection can be added onto the screen by pressing [DRAW] **7:Text**, moving the cursor near the point of intersect and typing your results.

    Note: *Text appears below and to the right of the cursor.*

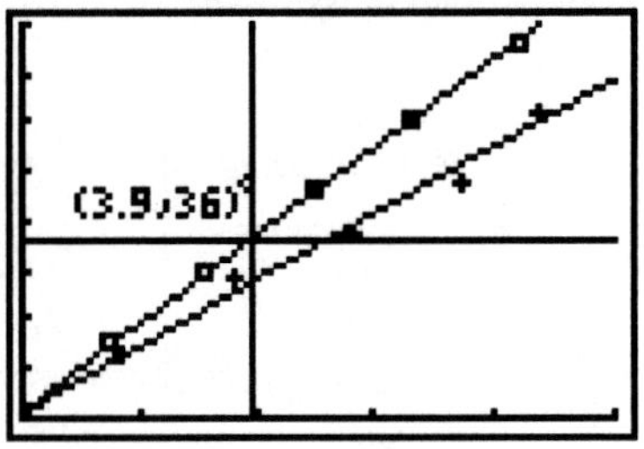

Predicting the mass of pennies

You can predict the mass of pennies if you know their volume.

1. Your teacher will give you 15 pennies and tell you whether they are pre-1982 or post-1984. Using the equation $D = \frac{M}{V}$ and the density determined earlier, measure their volume by water displacement and mathematically determine their mass.

2. Verify your answer using the TI-73. Press [TRACE] and press [▼] until you see the equation for **Y1** (if the pennies are pre-1982) or **Y2** (if the pennies are post-1984).

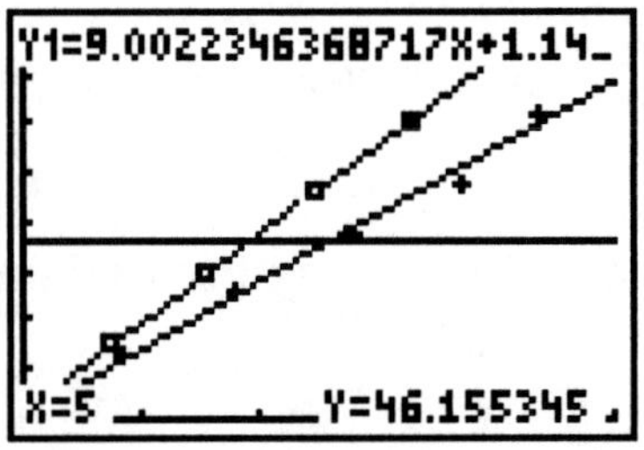

3. Type the volume of the pennies, and press [ENTER].

    Note: *3.7 was used for the value in this example.*

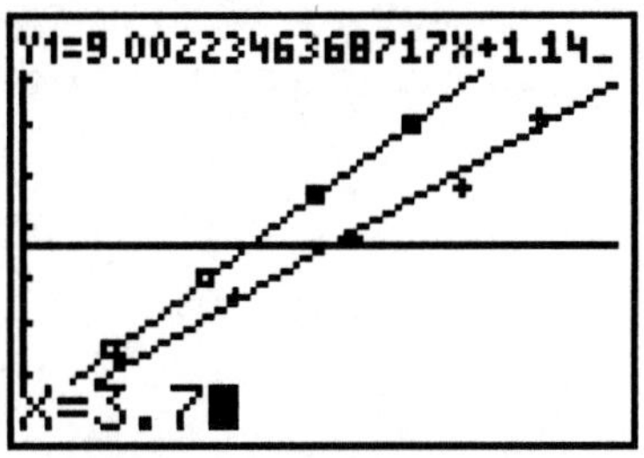

4. The *y* value in the lower right hand corner of the screen is the mass of the 15 pennies.

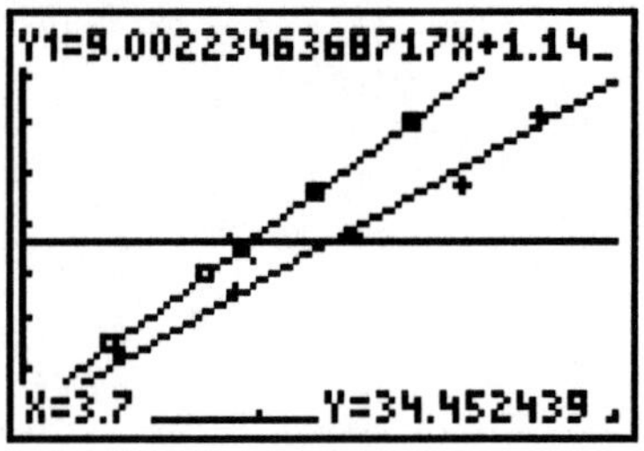

Data Collection and Analysis

Name ____________________

Date ____________________

Activity 6: You're So Dense

Collecting the data

Number of pennies	Pre-1982		Post-1984	
	Volume (cm^3)	Mass (g)	Volume (cm^3)	Mass (g)
0	0	0	0	0
5				
10				
15				
20				
25				

Analyzing the data

1. The *slope* of the linear regression line for pre-1982 pennies is:

   ____

2. The *slope* of the linear regression line for post-1984 pennies is:

   ____

3. Explain what the *slope* represents. For a linear model, is the slope the same along the entire model? Explain.

   ____

   ____

4. In theory, the equations should go through the origin. In other words, the *y-intercept* should be 0. The *slope* of a linear model is given by the equation shown here. The first point may be the origin, (0, 0). Therefore, the equation for the slope of *this* particular model can also be written as

$$A = \frac{Y_2 - Y_1}{X_2 - X_1}$$

$A = \frac{Y}{X}$ when Y and X are greater than 0.

Observe that Y is the mass and X is the volume, therefore

$Slope = \frac{mass}{volume}$.

What is another term for $\frac{mass}{volume}$?

____

5. Using information from questions 1 and 2, what is the density of the pre-1982 pennies and what is the density of the post-1984 pennies?

Density of pre-1982 pennies ____________

Density of post-1984 pennies ____________

Extensions

- Construct mass versus volume linear models as you did with the pre-1982 and post-1984 pennies, but use different liquids. Compare alcohol and water or compare water and oil.

  Determine the slope (density) for each liquid studied.

- Construct mass versus volume linear models as you did with the pre-1982 and post-1984 pennies, but compare bird bones and mammalian bones. Measure volume by water displacement as you did with the pennies.

  Determine the slope (density) for each bone type studied.

Teacher Notes

Activity 6

You're So Dense

Objectives

- To investigate the relationship between mass and volume
- To find the *x* value of a function, given the *y* value
- To find the *y* value of a function, given the *x* value
- To use technology to find a linear regression
- To use technology to plot a set of ordered pairs

Materials

- TI-73 graphing device
- 50 ml graduated cylinder
- Balance
- Pennies (25 pre-1982 and 25 post-1984)

Preparation

- The density of an object is an *intensive* property, which means that it is independent of the size of the sample. Any part of the object is representative of the whole, hence the linearity of the regression. Mass and volume by themselves are *extensive* properties, which means that they depend on size (that is, the larger an object, the larger its mass or volume).
- The density of an object changes if the temperature of the object changes. Generally, as an object cools, its density increases, but there are some important exceptions. One exception is water. The density of water increases as its temperature decreases to 4°C. As the temperature drops from 4°C to 0°C (the freezing point of water), the density decreases. Therefore, ice floats – a fact that has important biological significance.
- In 1982, the U.S. government changed the way pennies were minted. Up until that time, pennies were minted from a predominantly copper alloy. Now pennies are minted from a zinc alloy and finished with a thin copper coating. As a result, the two types of pennies have distinctly different densities.

  The change in metal composition took place over a couple of years, depending upon where the pennies were minted. To avoid confusion in this activity, pennies minted in the years 1982-1984 are not to be used.
- The **Extensions** section provides other interesting examples of density that could be used in this activity. Density of increasing amounts of liquids could be examined. Water is denser than cooking oil. Hence, cooking oil will form a

separate layer on top of water. Cooking oil is denser than alcohol and thus will form a separate layer below the alcohol.

- Another interesting comparison could be made using cleaned bones from a bird and a mammal. Bird bones are less dense to allow for flight. Use pork or beef bones and compare to similar bones in a chicken or turkey. Scrub the bones in a bleach solution, rinse, and let dry.

Answers to Data Collection and Analysis questions

Collecting the data

Sample data:

Number of pennies	Pre-1982		Post-1984	
	Volume (cm^3)	Mass (g)	Volume (cm^3)	Mass (g)
0	0	0	0	0
5	1.5	15.21	1.7	12.64
10	3.1	30.09	3.6	25.18
15	5.0	45.94	5.6	37.89
20	6.6	61.09	7.4	48.00
25	8.4	75.99	8.7	62.37

Analyzing the data

1. The *slope* of the linear regression line for pre-1982 pennies is:

Slope (pre-1982) = 9.00.

2. The *slope* of the linear regression line for post-1984 pennies is:

Slope (post-1984) = 6.85.

3. Explain what the *slope* represents. For a linear model, is the slope the same along the entire regression? Explain.

The slope is a ratio of the change in mass to the change in volume. The slope remains the same, regardless of the size of the object.

4. In theory, the equations should go through the origin. In other words, the *y*-intercept should be 0. The *slope* of a linear model is given by the equation shown here. The first point may be the origin, (0, 0). Therefore, the equation for the slope of *this* particular model can also be written as

$$A = \frac{Y_2 - Y_1}{X_2 - X_1}$$

$A = \frac{Y}{X}$ when Y and X are greater than 0.

Observe that *Y* is the mass and *X* is the volume, therefore

$$Slope = \frac{mass}{volume}.$$

What is another term for $\frac{mass}{volume}$?

The slope is equal to the mass divided by the volume; therefore, the slope is the density of the pennies.

5. Using information from questions 1 and 2, what is the density of the pre-1982 pennies and what is the density of the post-1984 pennies?

Density of pre-1982 pennies = 9.00 grams per cubic centimeter.

Density of post-1984 pennies = 6.85 grams per cubic centimeter.

EXPLORATIONS

Activity 7

Now You See It, Now You Don't

Objectives

- To study the relationship between age and near point accommodation
- To predict a person's age based on near point accommodation
- To use technology to study an exponential regression
- To use technology to create a box-and-whisker plot
- To use technology to create a histogram

Materials

- TI-73 graphing device
- Metric ruler or meter stick
- String length: 1.5 meters

Introduction

Have you ever noticed someone holding an object away from his or her face so that they can see it more easily? This may seem somewhat perplexing. Why would holding an object *further* away make it *easier* to examine?

Focusing one's eye when looking at a close object is referred to as *near point accommodation*. The focusing of one's eye is made possible by little muscles that pull on the eye, and therefore slightly reshape the lens. This causes the lens to focus the light on the sensitive cells of the retina.

Problem

Is near point accommodation related to a person's age?

Collecting the data

Follow steps 1 through 4 to determine a person's near point accommodation. Collect data at home using subjects in the 35 - 64 age group.

1. Have your subject hold the letter "**a**", shown at the right, in front of his/her face. Move the page as close to the subject as possible, keeping the letter in focus.

   a

   Note: *If your subject normally wears glasses, then the glasses should be worn during this test (unless the glasses are reading glasses, in which case the glasses should not be worn). If your subject wears bifocals, then the subject should wear his / her glasses and should view the letter through the upper part of the glasses that are used for distance. If your subject wears separate glasses for distance and reading, then he / she should wear the glasses that are used for distance.*

2. Use a string to determine the distance from the front of the eye to the letter "**a**". **Do not hold the string too close to the eye.** Holding the string along your subject's face, extend the string from the cornea (the front part of the eye) to the page with the letter "**a**". In the example shown here, the near point accommodation is 15 centimeters.

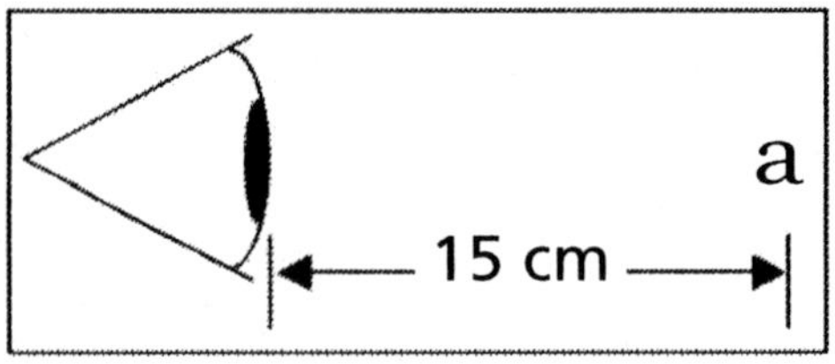

3. Record the result on the **Data Collection and Analysis** page. Be sure to indicate the age of the subject. Repeat this procedure with four other subjects.

4. Submit your data to your teacher. Your teacher will collate the data from all members of the class.

Setting up the TI-73

Before starting your data collection, make sure that the TI-73 has the STAT PLOTS turned OFF, Y= functions turned OFF or cleared, the MODE and FORMAT set to their defaults, and the lists cleared. See the Appendix for a detailed description of the general setup steps.

Entering the data in the TI-73

1. Press [LIST].

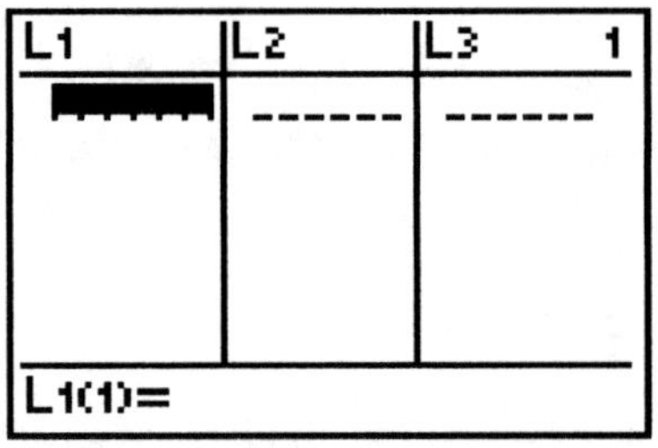

2. Enter the age of the subjects in **L1**.

3. Enter the near point accommodation of all the subjects in **L2**. (Make sure that the pairs of age and near point accommodation data match in each column.)

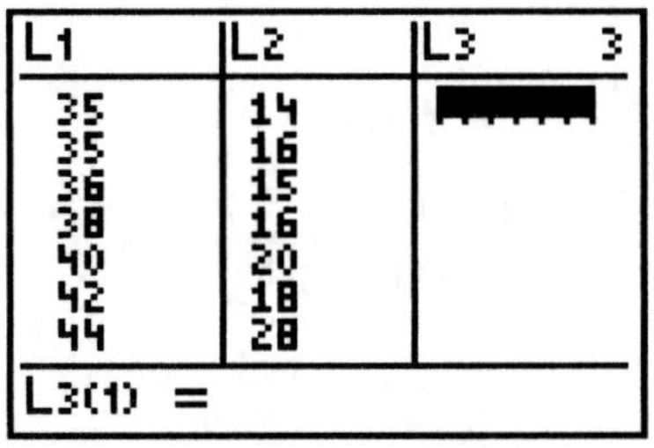

4. Enter the near point accommodation of only the 35 - 44 age group subjects in **L3**.

5. Enter the near point accommodation of only the 45 – 54 age group subjects in **L4**.

6. Enter the near point accommodation of only the 55 – 64 age group subjects in **L5**.

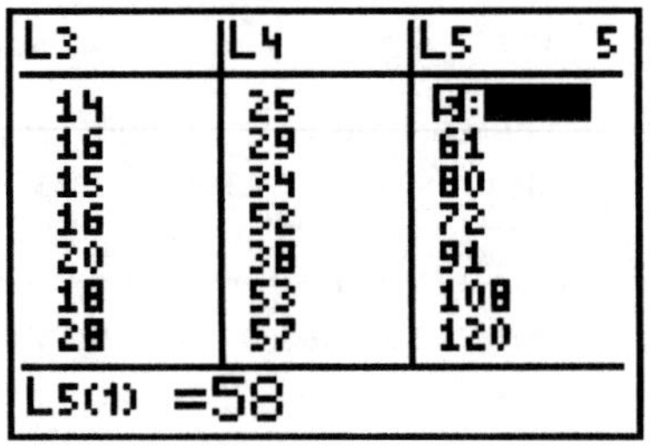

Setting up the window

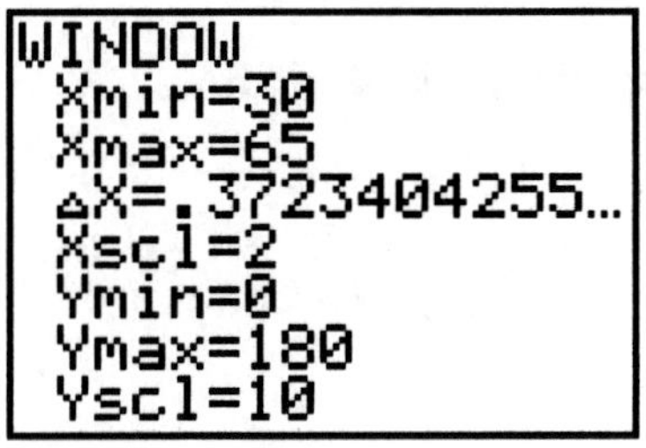

1. Press [WINDOW] to set up the proper scale for the axes.
2. Set the **Xmin** value by identifying the minimum value in **L1**. Choose a number that is less than the minimum.
3. Set the **Xmax** value by identifying the maximum value in each list. Choose a number that is greater than the maximum. **Do Not Change the ΔX Value.** Set the **Xscl** to **2**.
4. Set the **Ymin** value by identifying the minimum value in **L2**. Choose a number that is less than the minimum.
5. Set the **Ymax** value by identifying the maximum value in **L2**. Choose a number that is greater than the maximum. Set the **Yscl** to **10**.

Graphing the data: Setting up a scatter plot

You can analyze the data in several different ways. You will need to set up a scatter plot and model the data (exponential regression). You can then use the data collected to predict a person's age based on their near point accommodation.

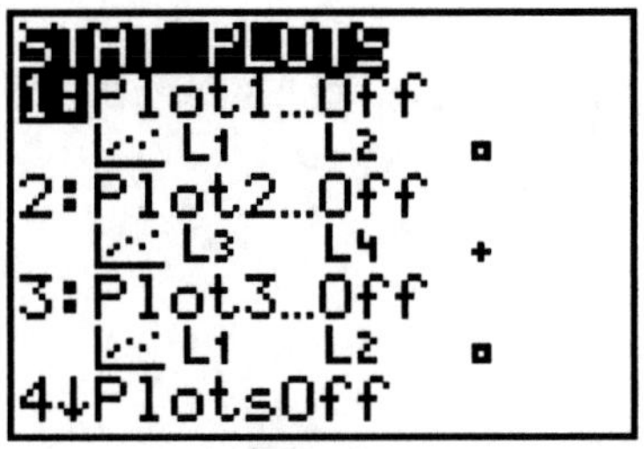

1. Press [2nd] [PLOT]. Select **1:Plot1** by pressing **1** or [ENTER].

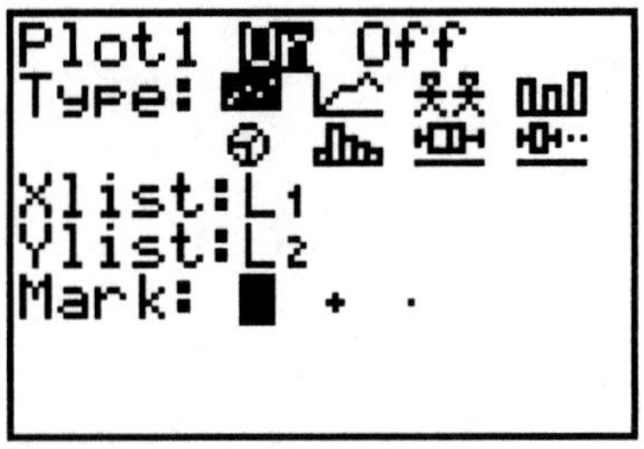

2. Set up the plot as shown by pressing [ENTER] [▼] [ENTER] [▼] [2nd] [STAT] **1:L1** [▼] [2nd] [STAT] **2:L2** [▼] [ENTER].

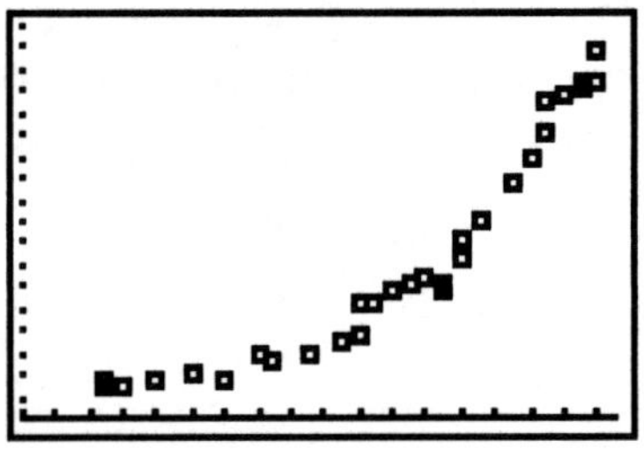

3. Press [GRAPH] to see the plot.

   Observe the shape of the plot. It is not linear because the slope changes. What type of regression would model such data?

Answer questions 1 and 2 on the **Data Collection and Analysis** page.

Analyzing the data

Finding a best fit line

1. Find an exponential model for the data. Press [2nd] [STAT] [◄] to move the cursor to the **CALC** menu.

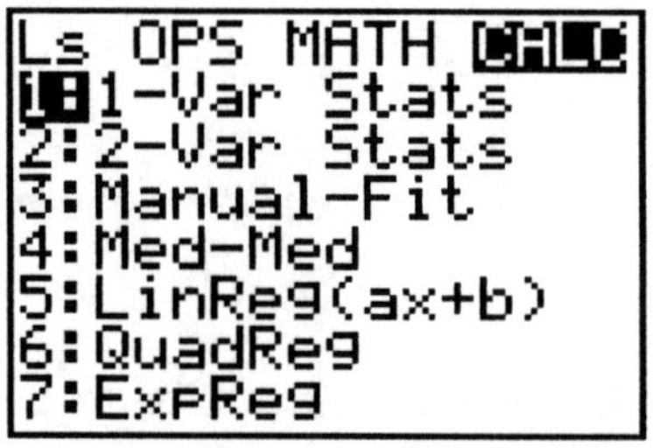

2. Select **7:ExpReg** by pressing **7**.

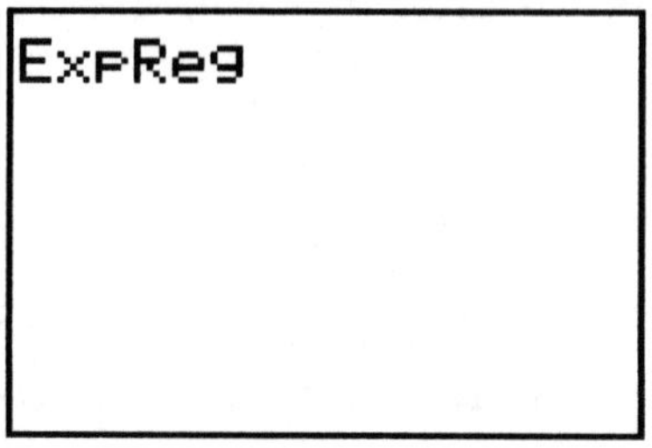

3. Press [2nd] [STAT] **1:L1** [,] [2nd] [STAT] **2:L2** [,].

4. Press [2nd] [VARS]. Select **2:Y-Vars** by pressing **2**.

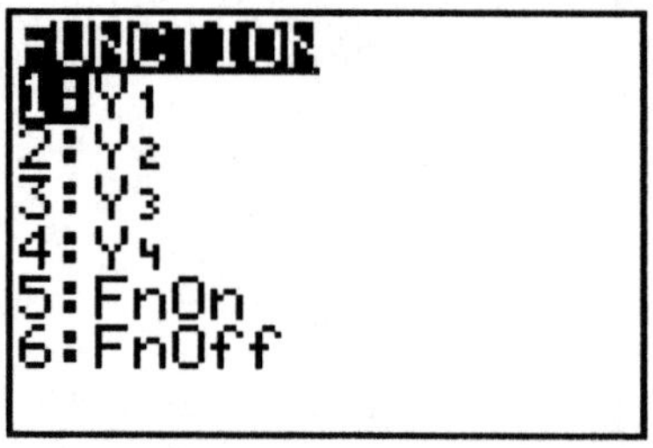

5. Select **1:Y1** by pressing **1** or [ENTER].

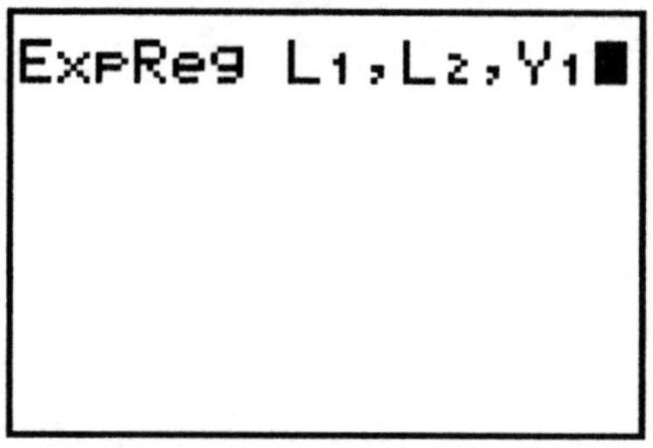

6. Press [ENTER] to calculate the exponential regression. The function is pasted in **Y1**.

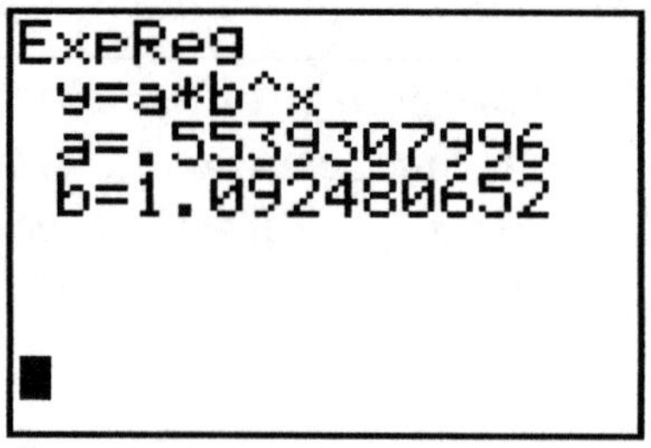

7. Press [GRAPH] to see the exponential regression model.

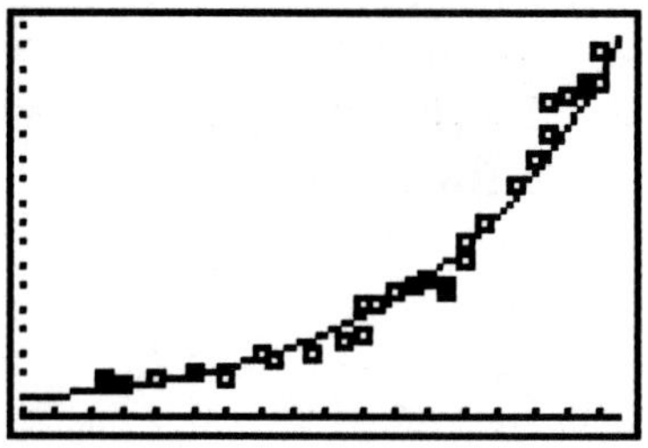

Answer questions 3 and 4 on the **Data Collection and Analysis** page.

Determining age based on the near point accommodation

If a person has a near point accommodation of 47 cm, how old is that person likely to be according to the data?

1. Press [Y=] and move the cursor to **Y2**. Enter **47**, the near point accommodation.

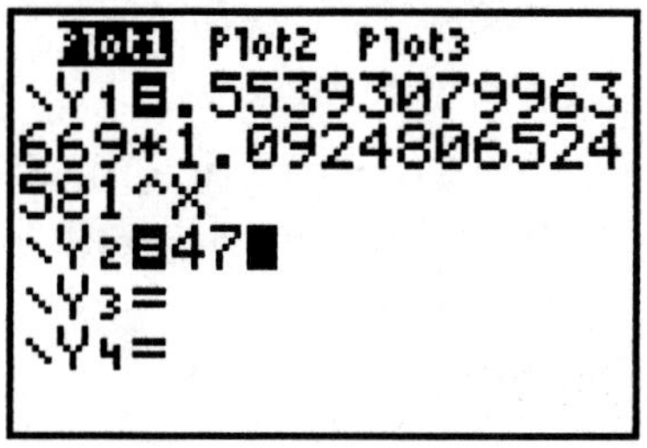

2. Press [GRAPH] to see the intersection of the two functions. The *x* value of the point where the two functions intersect is the predicted age of the person if their near point accommodation is 47 cm.

   The table function of the TI-73 will be used to determine the coordinates of the point of intersection.

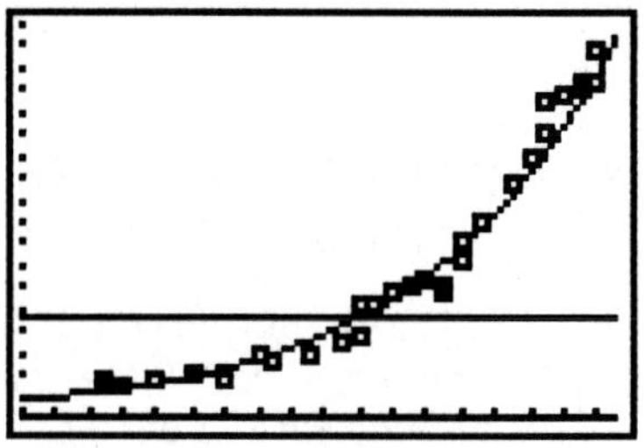

3. Press [2nd] [TBLSET]. Type the lowest value in **L1** (35 in the example). Press [▼] **5** to set the **ΔTbl** value.

4. Press [2nd] [TABLE]. If necessary, use [▼] and [▲] to scroll the table.

   Note: *For this example, in the **Y1** column, 47 cm falls between 46.147 and 71.814 which corresponds to ages 50 and 55. Based on this information, the table will be readjusted.*

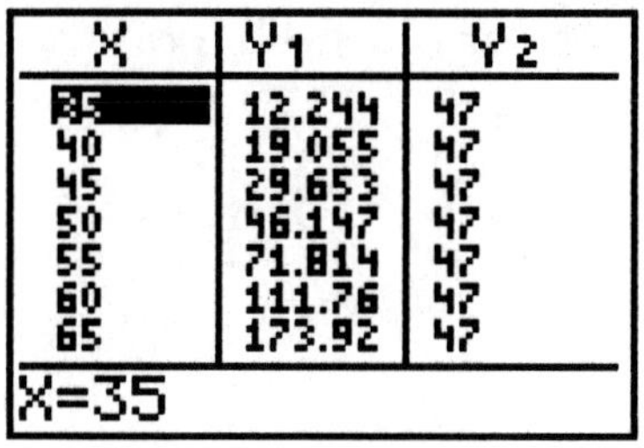

5. Press [2nd] [TBLSET]. Enter the result from Step 4 in **TblStart**. Press [▼] **0.1** to set the **ΔTbl** value.

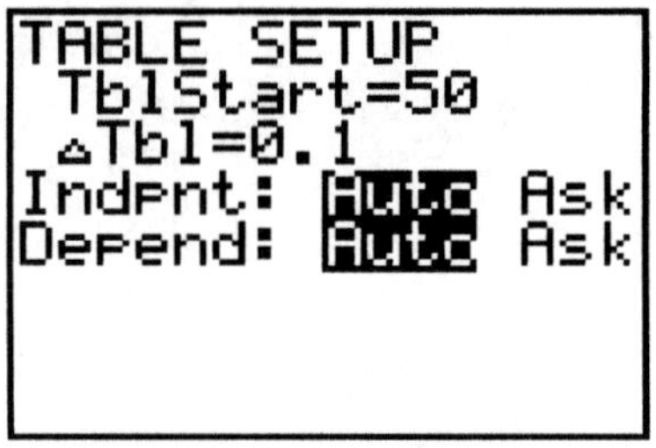

6. Press [2nd] [TABLE]. If necessary, use [▼] and [▲] to scroll the table.

   ***Note**: The data used to construct the exponential model used ages rounded to the nearest year. We should report age to the same level of precision. From the table, 47 falls between 46.971 and 47.388 which corresponds to 50.2 and 50.3. Rounding to the nearest year, the intersection point is (50, 47).*

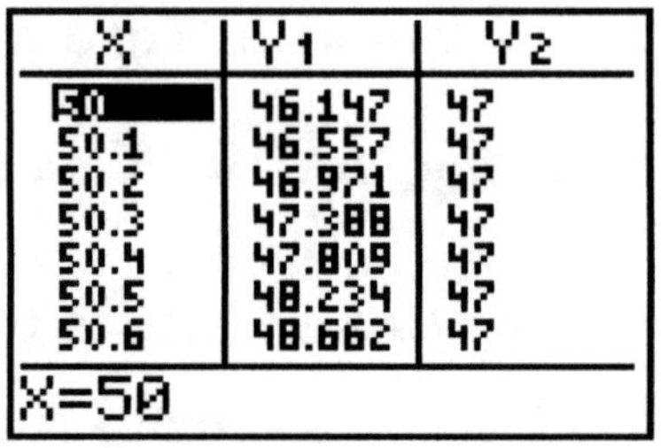

7. To verify this graphically, use the **DRAW** function. Press [DRAW]. Select **4:Vertical** by pressing **4**.

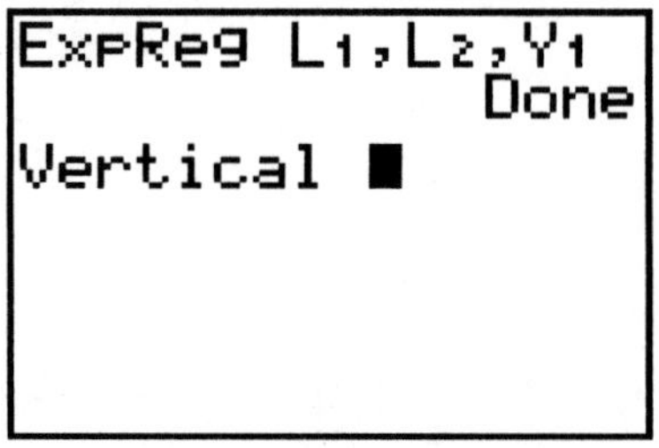

8. Type your results from Step 6 (for this example, **50**) and press [ENTER].

   Note: The coordinates of the point on the exponential model where all of the curves intersect is defined for this example by the vertical drawn at x=50 and the horizontal at y=47.

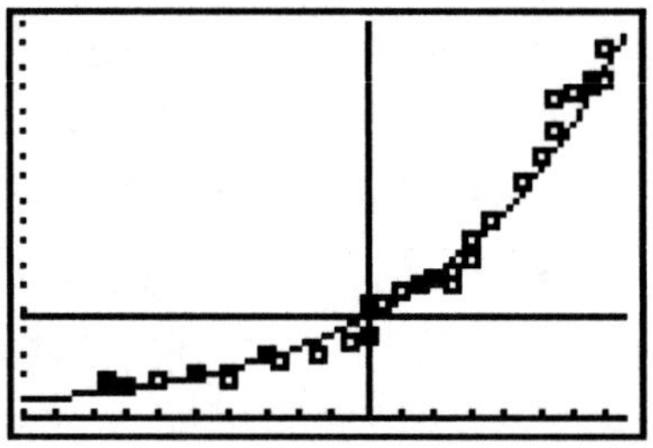

9. The coordinates of the intersection can be added onto the screen by pressing [DRAW] **7:Text**, moving the cursor near the point of intersect and typing your results.

   ***Note**: Text appears below and to the right of the cursor.*

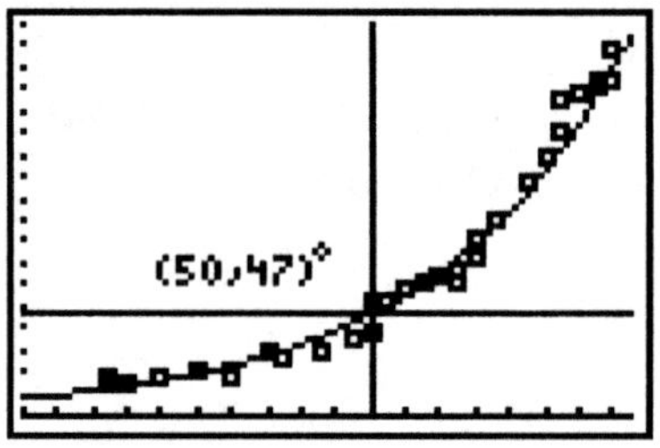

Answer question 5 on the **Data Collection and Analysis** page.

Graphing the data: Setting up a box-and-whisker plot

You can use a box-and-whisker plot to analyze the near point accommodation of the different age groups. You have already entered the near point accommodations of 35 - 44 year olds in **L3**, 45 - 54 year olds in **L4**, and 55 - 64 year olds in **L5**.

1. Press [2nd] [PLOT]. Select **1:Plot1** by pressing **1** or [ENTER].

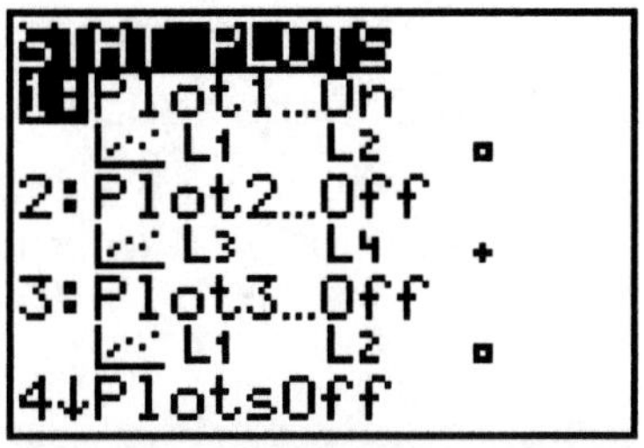

2. Set up the plot as shown by pressing [ENTER] [▼] [▶] [▶] [▶] [▶] [▶] [▶] [ENTER] [▼] [2nd] [STAT] **3:L3**. Press [▼] **1** to set the frequency.

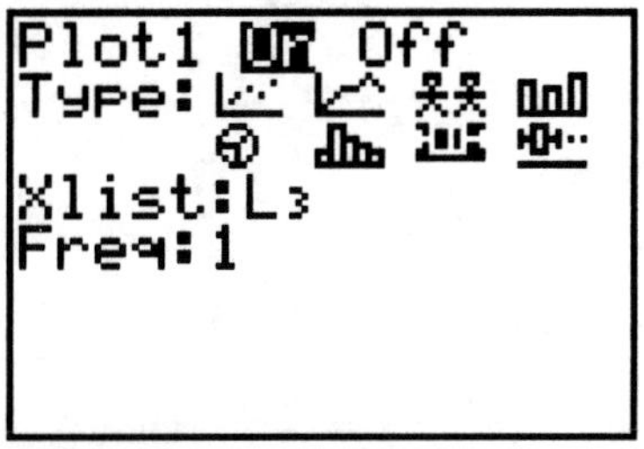

3. Press [2nd] [PLOT]. Select **2:Plot2** by pressing **2**.

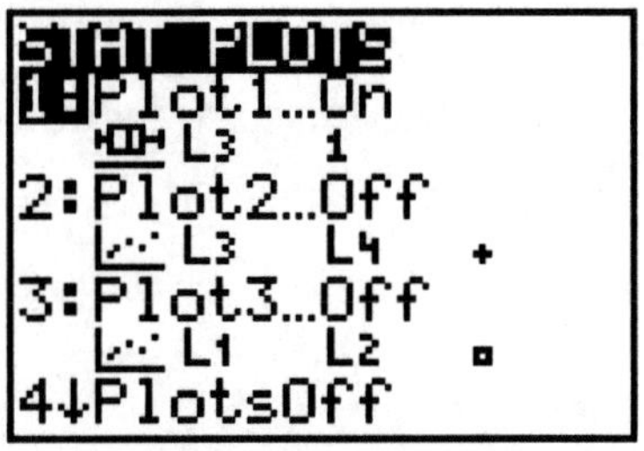

4. Set up the plot as shown by pressing [ENTER] [▼] [▶] [▶] [▶] [▶] [▶] [▶] [ENTER] [▼] [2nd] [STAT] **4:L4**. Press [▼] **1** to set the frequency.

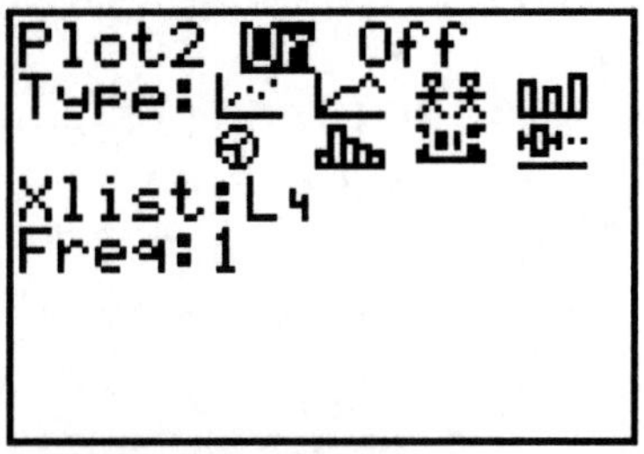

5. Press [2nd] [PLOT]. Select **3:Plot3** by pressing **3**.

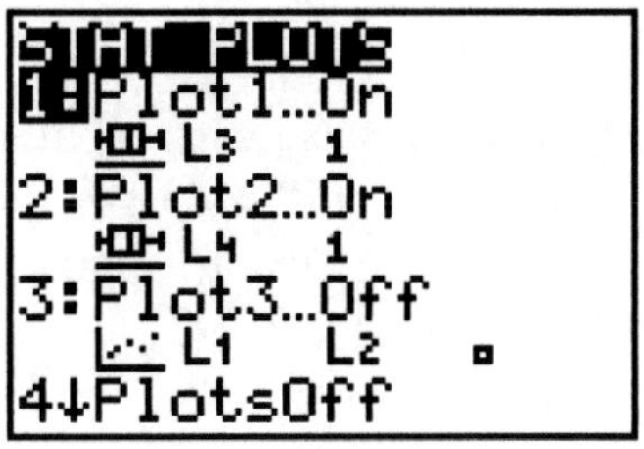

6. Set up the plot as shown by pressing [ENTER] [▼] [▶] [▶] [▶] [▶] [▶] [▶] [ENTER] [▼] [2nd] [STAT] **5:L5**. Press [▼] **1** to set the frequency.

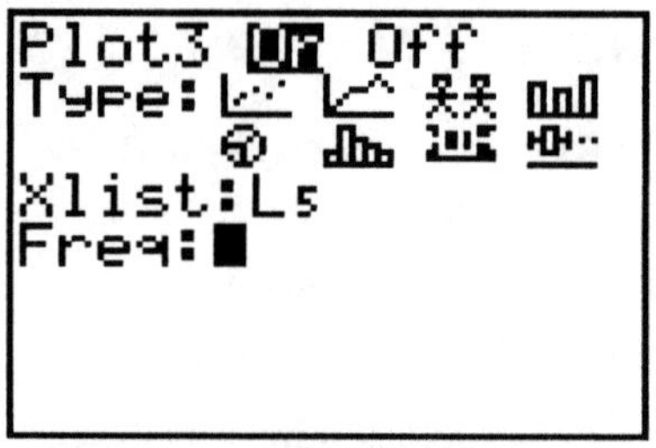

7. Turn OFF the equation in **Y1** and **Y2**. Press [Y=] [◀] [ENTER] [▼] [ENTER]. The equal signs should not be highlighted.

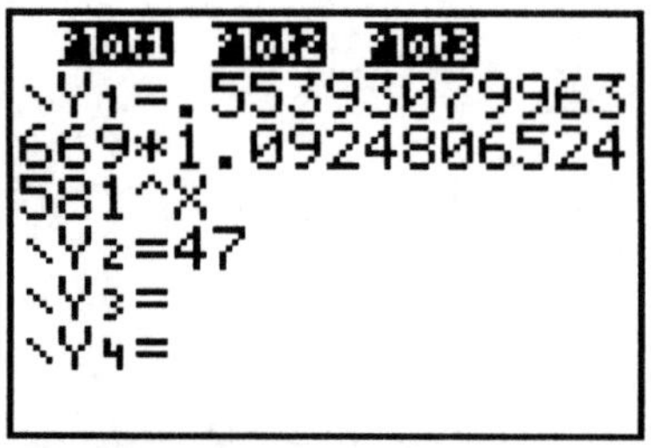

8. Press [ZOOM]. Select **7:ZoomStat** by pressing **7**.

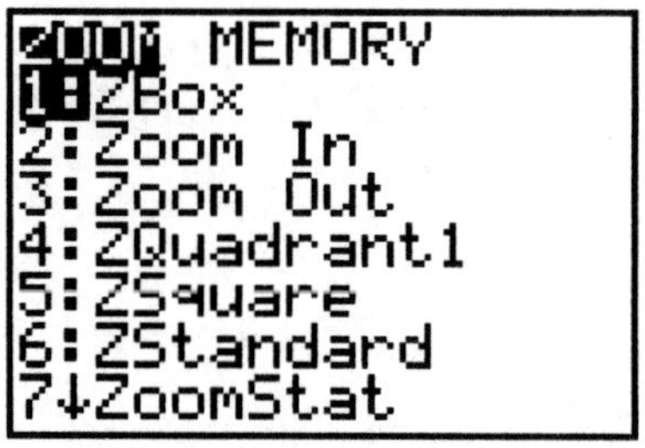

Observe the three box-and-whisker plots.

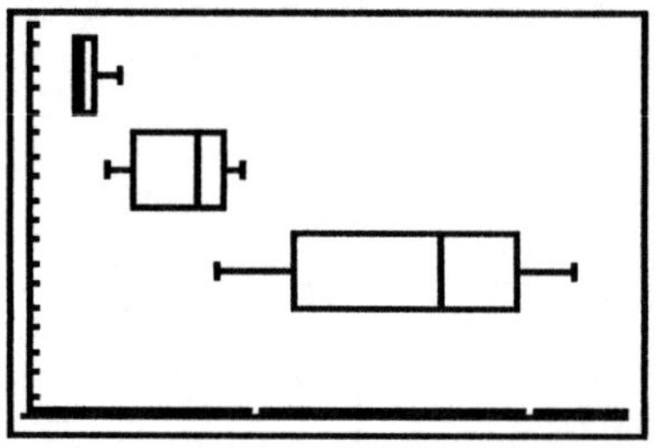

The box-and-whisker plots allow you to visualize the minimum, 1st quartile, median, 3rd quartile, and maximum values of a data list. You can easily make comparisons by viewing the plots of the three data sets.

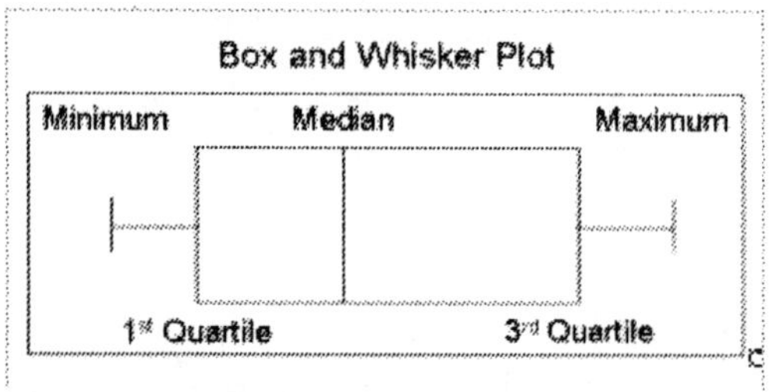

9. Press [TRACE]. Use [▼] and [▲] to move between plots. Use [◀] and [▶] to see the values of the median, quartiles, and extreme values.

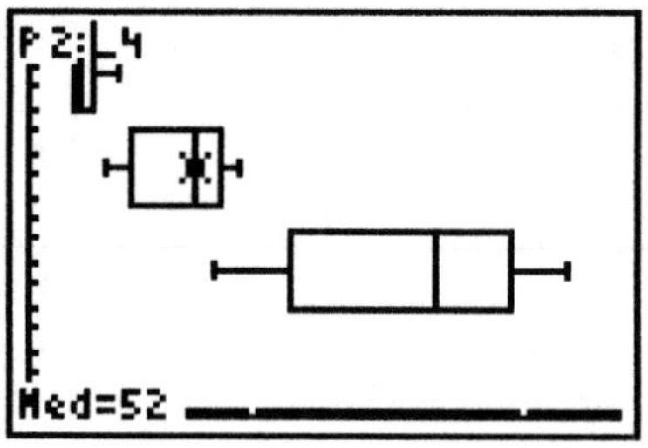

Examine these plots and provide the values for each plot on the **Data Collection and Analysis** page.

Answer questions 6 through 8 on the **Data Collection and Analysis** page.

Data Collection and Analysis

Name ______________________

Date ______________________

Activity 7: Now You See It, Now You Don't

Near Point Accommodation Data: Home Survey

Collecting the data

Age	Near Point Accommodation

Age Group	Minimum	1st Quartile	Median	3rd Quartile	Maximum
35 - 44 (**L3**)					
45 - 54 (**L4**)					
55 - 64 (**L5**)					

Analyzing the data

1. Describe the shape of the near point accommodation versus age plot. Is there a correlation between age and near point accommodation? Explain.

____

____

2. In which age group is the near point accommodation rising fastest: 35 - 44, 45 - 54, or 55 - 64? How can you tell by looking at the plot?

____

____

3. Write the exponential regression equation.

____

4. Does the exponential model seem to fit your data? Explain. (Does it seem to fit some age groups better than others?)

____

____

5. A person with a near point accommodation of 47 cm is likely to be ______ years old according to the exponential model.

6. What is the *median* near point accommodation of the three age groups analyzed?

   Ages 35-44: _________ Ages 45-54: _________ Ages 55-64: _________

7. Observe the width of the three box-and-whisker plots. Which plot has the largest width? Which plot has the smallest width? How do these widths relate to the exponential regression analyzed earlier? (**Hint:** Each box-and-whisker plot represents the same number of years – 10.)

____

____

____

8. Do the box-and-whisker plots overlap? What does this tell you about the near point accommodations of the three age groups that were analyzed?

____

____

Extension

Use a histogram to analyze the extent to which near point accommodation varies within your age group. The following table shows some sample data (nearest 0.1 cm).

7.1	6.5	9.3	6.2	8.7	9.0	7.7	8.1	8.3	7.8	8.9	8.8	6.9	9.1	7.5	8.3

Examine the distribution of near point accommodations in your class.

Which near point accommodations are least common in your age group? Most common?

How could you adjust the window to study different distribution ranges within your age group?

Teacher Notes

Activity 7

Now You See It, Now You Don't

Objectives

- To study the relationship between age and near point accommodation
- To predict a person's age based on near point accommodation
- To use technology to study an exponential regression
- To use technology to create a box-and-whisker plot
- To use technology to create a histogram

Materials

- TI-73 graphing device
- Metric ruler or meter stick
- String length: 1.5 meters

Preparation

- You may choose to use the near point accommodation data provided. If you decide to have your class collect the data, there are several ways to collate it. One option is to type the near point accommodation data directly into the TI-73 in **L1** and **L2** and then share the data with the students by linking calculators. You may want to print the data on a handout or put it on the chalkboard and have the students copy it into their lists.
- Real studies relating age and near point accommodation (presbyopia) show a general increase with age, but the relationship is not neatly modeled mathematically. From ages 35 - 64, it is close to an exponential model.

Answers to Data Collection and Analysis questions

Collecting the data

Sample data:

Age (years) – L1	Near point (cm) – L2	Age (years) – L1	Near point (cm) – L2
35	14	54	65
35	16	55	58
36	15	55	61
38	16	56	80
40	20	56	72
42	18	57	91
44	28	59	108
45	25	60	120
47	29	61	146
49	34	61	132
50	52	62	147
50	38	63	150
51	53	63	155
52	57	64	167
53	62	64	153

Analyzing the data

1. Describe the shape of the near point accommodation versus age plot. Is there a correlation between age and near point accommodation? Explain.

   The data will vary, but generally, one finds that it rises slowly at first and then more steeply in the upper age groups. As one's age rises, the near point accommodation rises, somewhat exponentially.

2. In which age group is the near point accommodation rising fastest: 35 - 44, 45 – 54, or 55 - 64? How can you tell by looking at the plot?

   One would expect the steepest rise to be in the 55 - 64 age group. This is where the plot is the steepest.

3. Write the exponential regression equation.

   Answers will vary depending on what data is used. For the given data, the regression is:

   $Y = (0.56)\,(1.1)^X$

4. Does the exponential regression seem to fit your data? Explain. (Does it seem to fit some age groups better than others?)

Answers will vary depending on what data is used. Based on the data provided, the regression appears to fit the data, but clearly there are individuals that are off the regression line.

5. A person with a near point accommodation of 47 cm is likely to be ____ years old according to the exponential model.

Answers will vary depending on the data used. According to the data provided, the age of the individual is approximately 50. However, if one examines the actual data, it is obvious that there is a fair amount of variation in near point accommodations at any given age.

6. What is the *median* near point accommodation of the three age groups analyzed?

Answers will vary. For the sample data:

Ages 35 - 44: Median is 16 cm.
Ages 45 - 54: Median is 52 cm.
Ages 55 - 64: Median is 126 cm.

7. Observe the width of the three *box-and-whisker* plots. Which plot has the largest width? Which plot has the smallest width? How do these widths relate to the exponential model analyzed earlier? (**Hint:** Each box-and-whisker plot represents the same number of years – 10.)

The box-and-whisker plot is for a 10-year range. The higher the age, the wider the plot because in an exponential model for near point accommodation versus age, as the age increases, the near point accommodation values increase at a faster rate. For that age range, there is a greater amount of variation in near point accommodation values.

8. Do the box-and-whisker plots overlap? What does this tell you about the near point accommodations of the three age groups that were analyzed?

The box-and-whisker plots will probably overlap, showing that there are subjects in one age group that have higher near point accommodations than subjects in the next older age group.

EXPLORATIONS

Activity 8

At a Snail's Pace

Objectives

- To plot a mathematical relationship that defines a spiral
- To use technology to create a spiral similar to that found in a snail
- To use technology to plot a set of ordered pairs

Materials

- TI-73 graphing device
- Lead pencil
- Colored pencils - red and green
- Metric ruler
- Compass
- Tracing paper and graph paper

Introduction

It is hard to imagine a way to mathematically describe a spiral. Spirals are commonly seen in nature. There are a number of different types of spirals. Some interesting examples include the spiral arrangements of scales in pine cones, seeds in a sunflower, or the calcified layers of a snail's shell. One way of constructing a spiral is to use the Fibonacci sequence. This sequence was examined in *Activity 4: The Calcumites Are Coming!* The sequence starts with the following series of numbers: 0, 1, 1, 2, 3, 5, 8, 13, 21, 34, 55, 89, and so on. Starting with the third number, each number is equal to the sum of the two numbers preceding it. This sequence can be applied to constructing a spiral resembling a snail's shell.

Problem

A snail shell is composed of a spiral of calcified layers that get wider with each turn. How can this spiral be mathematically modeled?

Collecting the data

1. Place a piece of graph paper on your desk. Draw a set of *x*- and *y*-axes to include the four extreme points (0, 0), (34, 0), (0, 21), and (34, 21).
2. Draw a square by connecting the vertices (10, 5), (10, 6), (9, 6), and (9, 5). Lightly shade this square with a pencil.

3. Construct a square to the left of the original square by connecting vertices (8, 5), (8, 6), (9, 6), and (9, 5). Observe that the two squares together form a rectangle whose width is 1 and whose length is 2.

4. Construct a square above this rectangle by connecting the vertices (10, 6), (10, 8), (8, 8), and (8, 6).

5. Continue to add squares to the graph in a clockwise fashion (see diagram at right) connecting the vertices shown in the table below. The two rows are the 1 X 1 and 2 X 2 squares that you already constructed. The last row is a 21 X 21 square.

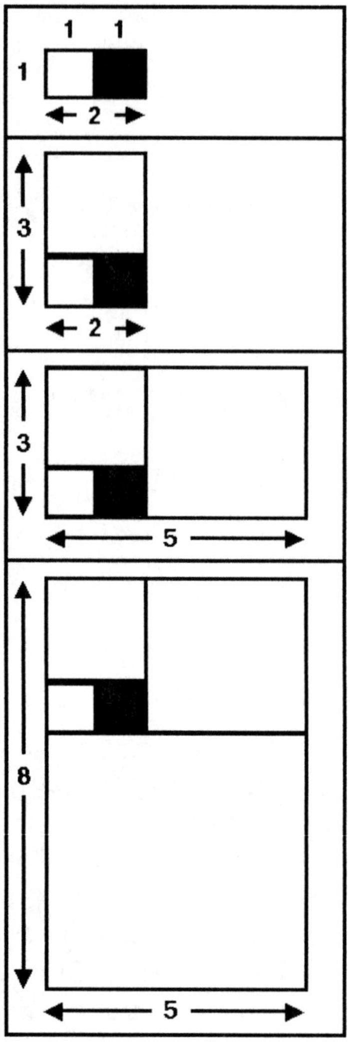

Length of side of square	Vertices			
1	(8, 5)	(8, 6)	(9, 5)	(9, 6)
2	(8, 6)	(8, 8)	(10, 6)	(10, 8)
3	(10, 5)	(10, 8)	(13, 5)	(13, 8)
5	(8, 0)	(8, 5)	(13, 0)	(13, 5)
8	(0, 0)	(0, 8)	(8, 0)	(8, 8)
13	(0, 8)	(0, 21)	(13, 8)	(13, 21)
21	(13, 0)	(13, 21)	(0, 34)	(34, 21)

6. Each time a square is added a new rectangle is formed. Record the dimensions of each rectangle on the **Data Collection and Analysis** page.

Answer questions 1 and 2 on the **Data Collection and Analysis** page.

7. Using a red pencil, mark each of the points shown in the table below on your graph paper. These points are each one of the corners of the squares in the graph. The dimensions of the sides of each square are also given in the table.

Point	(9, 6)	(10, 6)	(10, 5)	(8, 5)	(8, 8)	(13, 8)	(13, 0)	(0, 0)
Dimension of sides	1	2	3	5	8	13	21	34

8. Using a compass, construct a ¼ circle in the square that measures 1 X 1. Place the point of the compass on the red point at vertices (9, 6) that you made in step 7. Open up the compass so that the pencil is on an adjacent vertex. Sweep an arc that goes to the opposite corner of the square.

9. Repeat this step with each of the squares, using the red marked points as the place to position the compass point. Observe that you are constructing a continuous curve that resembles a spiral — a Fibonacci spiral. The result of the first three squares is shown to the right.

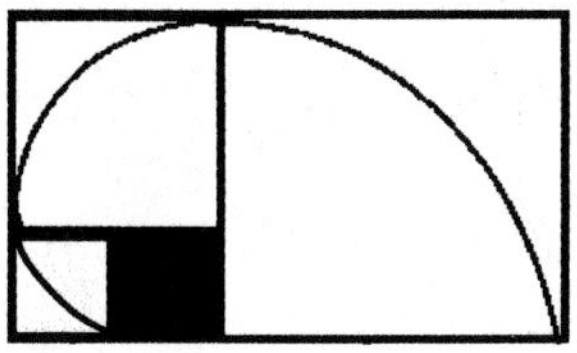

Setting up the TI-73

Before starting your data collection, make sure that the TI-73 has the STAT PLOTS turned OFF, Y= functions turned OFF or cleared, the MODE and FORMAT set to their defaults, and the lists cleared. See the Appendix for a detailed description of the general setup steps.

Setting up the window

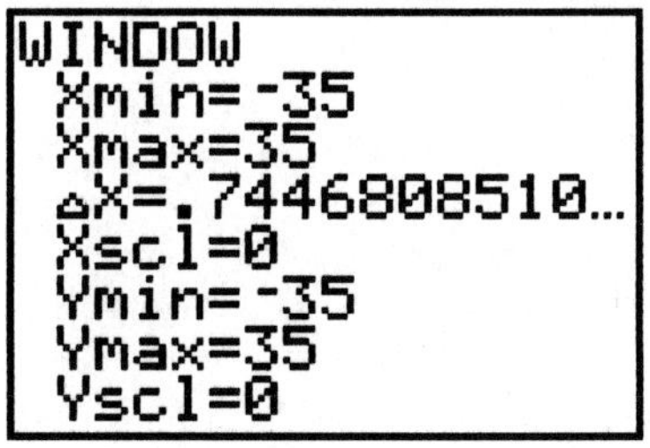

1. Press [WINDOW] to set up the proper scale for the axes.

2. Set the **Xmin** value by identifying the minimum value. Choose a number that is less than the minimum.

3. Set the **Xmax** value by identifying the maximum value in each list. Choose a number that is greater than the maximum. **Do Not Change the ΔX Value.** Set the **Xscl** to **0**.

4. Set the **Ymin** value by identifying the minimum value in **L2**. Choose a number that is less than the minimum.

5. Set the **Ymax** value by identifying the maximum value in **L2**. Choose a number that is greater than the maximum. Set the **Yscl** to **0**.

Entering the data in the TI-73

Construct a Fibonacci spiral using the TI-73. You can construct circles using the same center points and radii that were used in the previous steps. Although you must draw full circles, you are only interested in one-fourth of each circle. Collectively, the circle segments will form the Fibonacci spiral.

1. Press [2nd] [QUIT] to return to the Home screen. Press [CLEAR] to clear the Home screen.

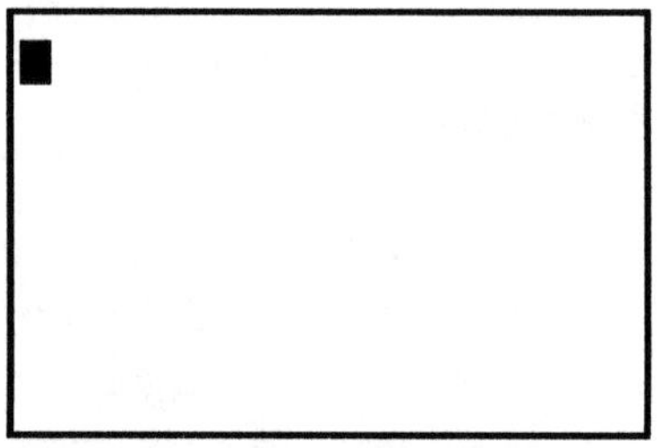

2. The first circle has a center at point (9, 6) and a radius of 1. Press [DRAW] to draw the circle.

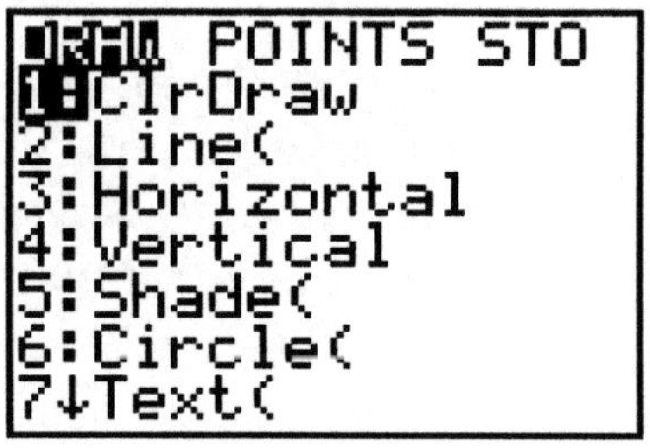

3. Select **6:Circle(** by pressing **6**.

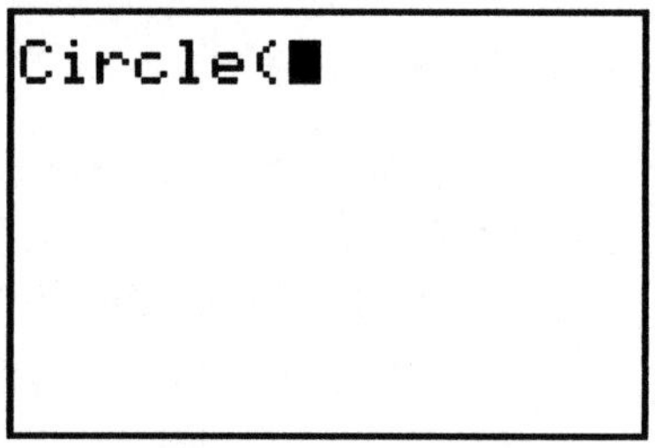

4. Type **9** [,] **6** [,] **1** [)].

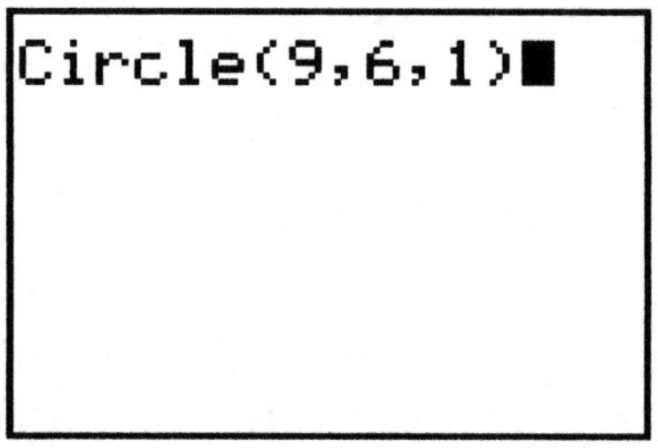

5. Press [ENTER] to draw the circle. A small circle is drawn as shown. The center of the circle is (9, 6) and the radius is 1.

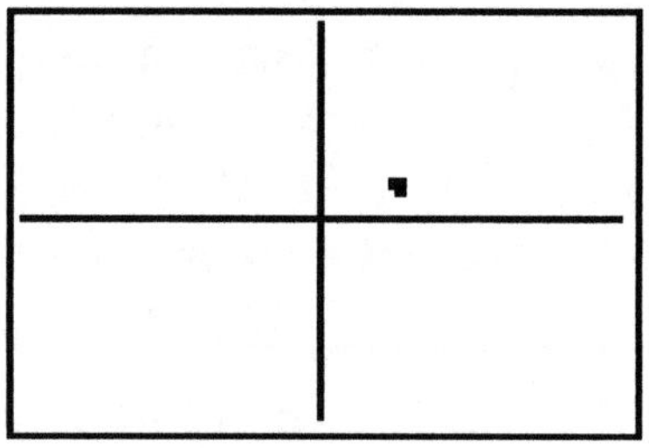

6. Repeat steps 1 through 4 for each circle. Change the center point and radius according to the table below each time step 4 is repeated.

Center point	Radius	Home screen shows
(10, 6)	2	Circle(10,6,2)
(10, 5)	3	Circle(10,5,3)
(8, 5)	5	Circle(8,5,5)
(8, 8)	8	Circle(8,8,8)
(13, 8)	13	Circle(13,8,13)
(13, 0)	21	Circle(13,0,21)
(0, 0)	34	Circle(0,0,34)

Where is the spiral on the diagram shown at the right?

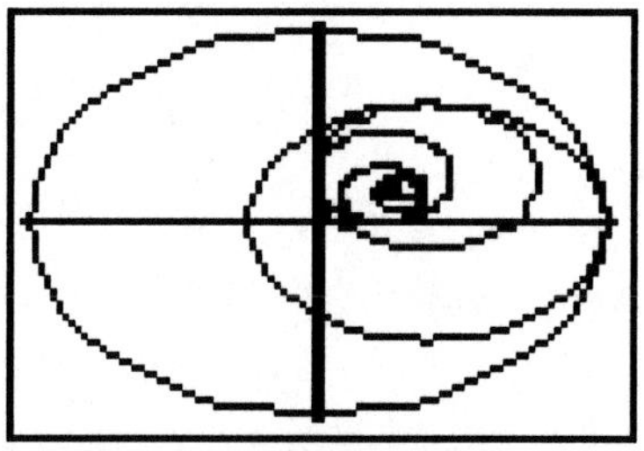

The first circle that you constructed appears as a dot in the first quadrant of the graph. Remember that you only need one-fourth of each circle that was drawn.

In the diagram shown at the right, three-fourths of each circle that was constructed is erased, leaving a spiral. This action cannot be done on the TI-73.

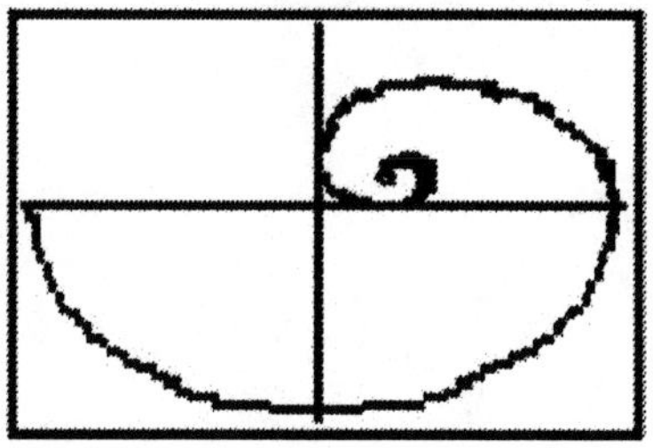

7. Use tracing paper to help spot the spiral on the TI-73. Using a pencil, draw a set of axes on the tracing paper. Place the paper over the TI-73 screen so that the axes on the paper align with the axes on the screen. Using a red pencil, find and trace the spiral by finding the dark center of the spiral and proceeding clockwise around it.

8. Use the prior screens as guidelines. You should now have a spiral. How does this spiral compare with the spiral that you constructed on the graph paper?

9. Save the drawing. Press [DRAW] [◄] to move the cursor to the **STO** menu.

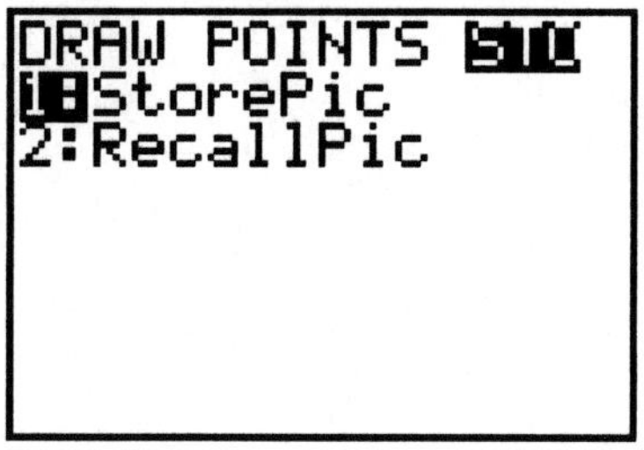

10. Select **1:StorePic** by pressing **1** or [ENTER].

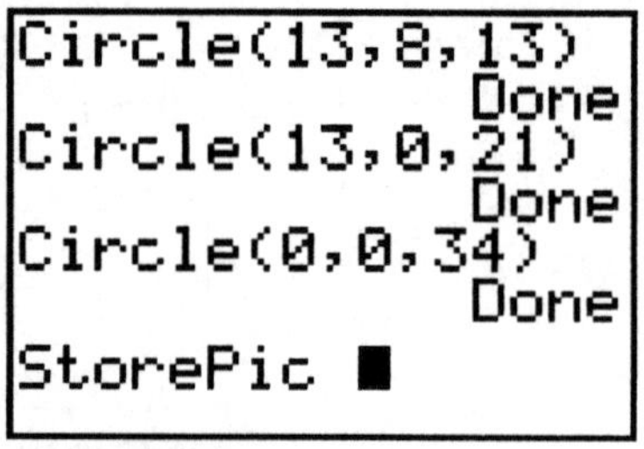

11. Press [2nd] [VARS]. Select **4:Picture** by pressing **4**.

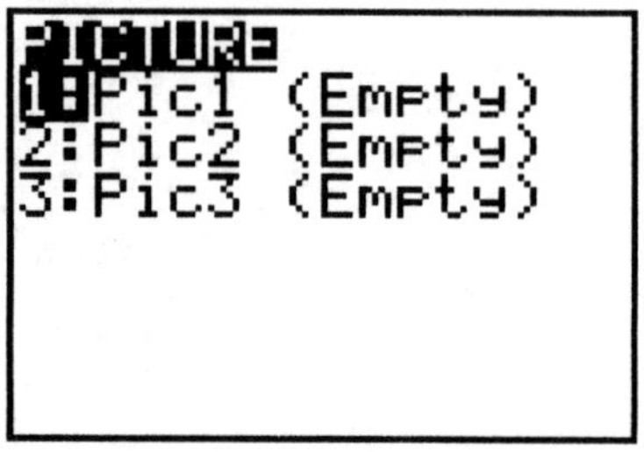

12. Select **1:Pic1** by pressing **1** or [ENTER].

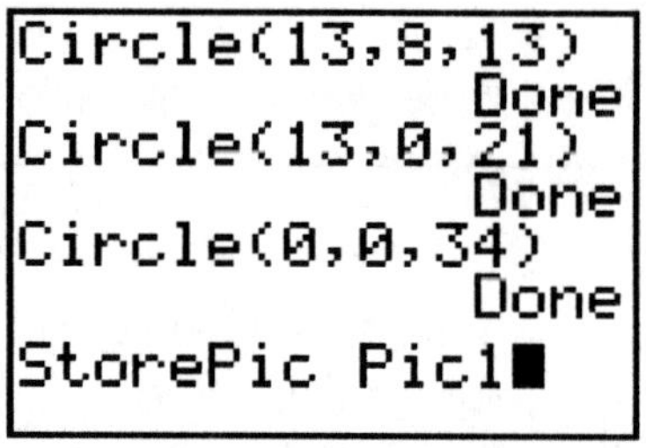

13. Press [ENTER] to save the picture for future use.

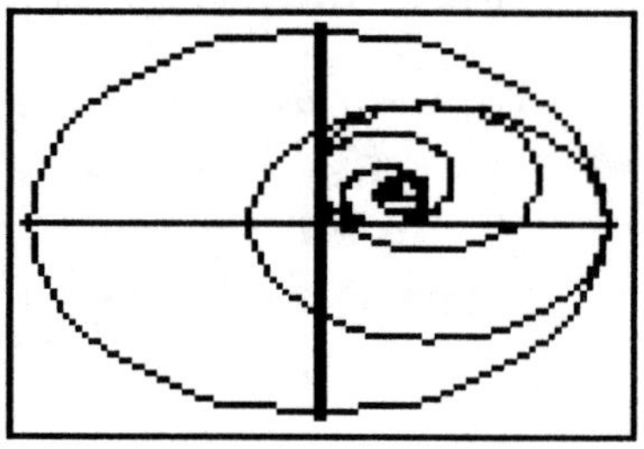

Data Collection and Analysis

Name ____________________

Date ____________________

Activity 8: At a Snail's Pace

Collecting the data

Dimensions of Rectangles	
Width	**Length**
1	2

Analyzing the data

1. Each square was added in a clockwise direction around the original shaded square. How many squares have to be added to the plot in order to draw one complete turn of the spiral?

   ____

2. The sequence of widths and lengths in the table above are part of a Fibonacci sequence. Determine the next five widths and lengths. Add those to the table below.

Width	Length

3. Give examples of Fibonacci sequences from this activity.

____

____

Extension

How much do the Fibonacci rectangles increase in area each time a square is added on?

Look at how the areas of the rectangles change as you build larger and larger rectangles. Find out by what factor the area changes each time you increase the size of the rectangles. Two formulas can be used:

$L - S = \Delta A$

$\Delta A \div S =$ Increase Factor

where *L* is the area of the large rectangle, *S* is the area of the small rectangle, and ΔA is the change in area.

Is there a golden lesson in this activity? Does the graph generated in this exercise support your ideas about a Golden Ratio?

Note: *Activity 4: The Calcumites Are Coming! and this activity both dealt with the Golden Ratio of 1.618. This ratio is obtained by dividing the numbers in a Fibonacci sequence by the preceding values in the sequence. Observe that the area of the rectangles increases by a factor of 1.618. When using a spiral of squares to construct a spiral each quarter turn of the spiral increased in distance from the center by a factor of 1.618. In a Fibonacci spiral, the distance from the center increases by a factor of 1.618 with each complete turn of the spiral.*

Teacher Notes

Activity 8

At a Snail's Pace

Objectives

- To plot a mathematical relationship that defines a spiral
- To use technology to create a spiral similar to that found in a snail
- To use technology to plot a set of ordered pairs

Materials

- TI-73 graphing device
- Lead pencil
- Colored pencils - red and green
- Metric ruler
- Compass
- Tracing paper and graph paper

Preparation

- For the first part of this activity it is important to select appropriate graph paper that allows students to visualize the spiral of squares and the spiral that they generate when drawing the arcs within the squares.
- Fibonacci sequences can be seen throughout this activity: 1) sides of the squares; 2) widths of the rectangles; 3) lengths of the rectangles; and 4) radii of the concentric circles that form the spiral.
- Golden Ratios are embedded within this activity, although only one is calculated. The one that the students calculated is the increase in the area of the Fibonacci rectangles. Other examples include: 1) the increase in distance from the center to subsequent turns in a Fibonacci spiral, and 2) the length of the rectangles divided by the width of the rectangles.
- The Fibonacci sequence is not the only sequence that approaches the Golden Ratio. Students can build a sequence by starting with any two numbers where, beginning with the third number, each number is equal to the sum of the preceding two numbers: 3, 7, 10, 17, 27, 44, 71, 115, and so on. The Golden Ratio is obtained by dividing each number by the preceding number in the sequence.

Answers to Data Collection and Analysis questions

Collecting the data

Sample data:

Dimensions of Rectangles	
Width	**Length**
1	2
2	3
3	5
5	8
8	13
13	21
21	34
34	55
55	89

Analyzing the data

1. Each square was added in a clockwise direction around the original shaded square. How many squares have to be added to the plot in order to draw one complete turn of the spiral?

   Four. Each square gives one-fourth of a turn of the spiral.

2. The sequence of widths and lengths in the table above are part of a Fibonacci sequence. Determine the next five widths and lengths. Add those to the table below.

   See chart.

Width	Length
89	144
144	233
233	377
377	610
610	987

3. Give examples of Fibonacci sequences from this activity.

   Fibonacci sequences can be seen throughout this activity: 1) sides of the squares; 2) widths of the rectangles; 3) lengths of the rectangles; and 4) radii of the concentric circles that form the spiral.

EXPLORATIONS

Activity 9

You're Probably Right, It's Wrong

Objectives

- To use technology to find experimental and theoretical probabilities
- To use technology to find measures of central tendencies
- To use technology to explore simulation
- To use technology to generate random numbers
- To use technology to plot a histogram
- To use technology to plot a pie chart
- To use technology to plot a pictograph
- To use technology to plot a bar graph

Materials

- TI-73 graphing device

Introduction

Nathan had a choice between studying for a mathematics test and going to the movies with a friend. He knew going to the movies was the wrong choice, but he decided to go anyway. When the math test was handed out the next day, he knew he should have studied. After seeing the test, it was clear that he was not prepared to take it. Nathan was somewhat relieved when he saw that the test had 20 multiple-choice questions. He knew that if he guessed the answers, he would have a 25% chance of getting the correct answer for each question, since each question had four choices. Nathan remembered that his TI-73 had a random number generator. He used this feature to help him guess the answers on the test. Nathan is now nervous about the results of the math test. If he fails this test, he will be grounded for a month. Nathan thinks that he did not pass the test. Is he right?

You will find the *experimental probability* to determine the likelihood that Nathan has passed this test. You will perform a *simulation* to determine the experimental probability. *Probability* is a number between 0 and 1 that measures the likelihood that an event will or will not occur. If the probability is *0*, then the probability that the event will occur is impossible. If the probability is *1*, then the probability that the event will occur is certain. *Experimental probability* is determined by performing experiments and observing outcomes to determine what might happen in a given situation. A *simulation* is a method for finding experimental probability using a device to model the event.

You will also find the *theoretical probability* of Nathan passing the test. If $P(E)$ represents the probability of the event occurring, m represents successful outcomes, and n represents possible equally likely outcomes (both successful and unsuccessful), then

$P(E) = \frac{m}{n}$ is the *theoretical probability* of the event occurring.

Problem

Was Nathan's idea of generating random numbers to answer the questions on the test a good idea? Should Nathan prepare to clean his room since he might be spending a good deal of time in there?

Collecting the data – Part I

Use the TI-73's random number generator to perform a simulation to guess answers on the test. The choices of answers are ***A***, ***B***, ***C***, or ***D***. An ***A*** will be represented by a **1**, a ***B*** by a **2**, a ***C*** by a **3**, and a ***D*** by a **4**. The correct answers for the test are listed below along with the corresponding number for the letter.

1. C – 3	**6.** C – 3	**11.** B – 2	**16.** C – 3
2. B – 2	**7.** A – 1	**12.** A – 1	**17.** D – 4
3. C – 3	**8.** B – 2	**13.** A – 1	**18.** D – 4
4. D – 4	**9.** D – 4	**14.** D – 4	**19.** C – 3
5. D – 4	**10.** C – 3	**15.** D – 4	**20.** A – 1

Setting up the TI-73

Before starting your data collection, make sure that the TI-73 has the STAT PLOTS turned OFF, Y= functions turned OFF or cleared, the MODE and FORMAT set to their defaults, and the lists cleared. See the Appendix for a detailed description of the general setup steps.

Entering the data in the TI-73

1. Press [LIST] and enter the data for the answers to the test in **L1**. When finished, press [2nd] [QUIT] to exit the list editor.

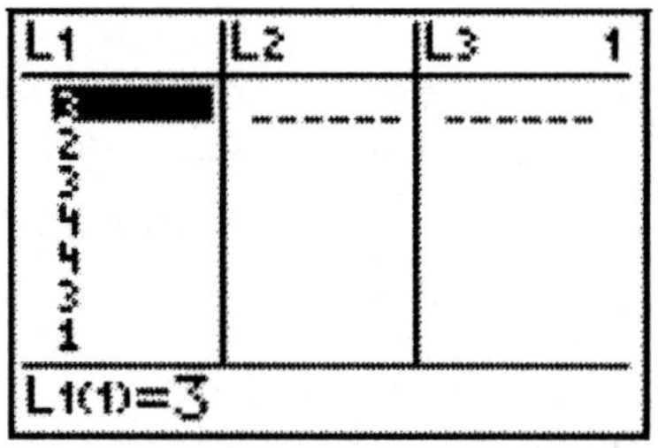

The following steps will generate a list of random numbers between 1 and 4 and store them in **L2**.

2. Press [MATH].

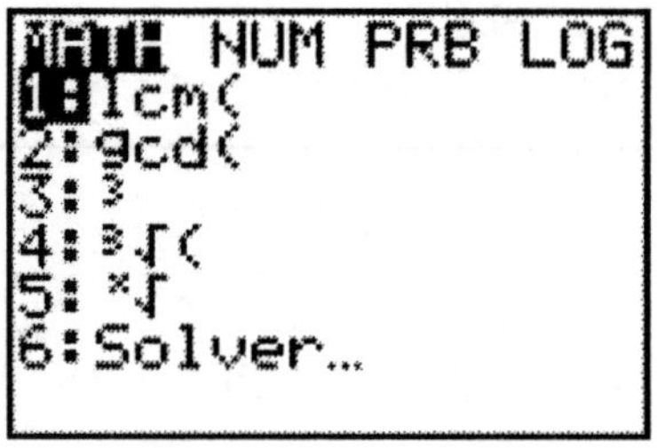

3. Press [▶] [▶] to move the cursor to the **PRB** menu.

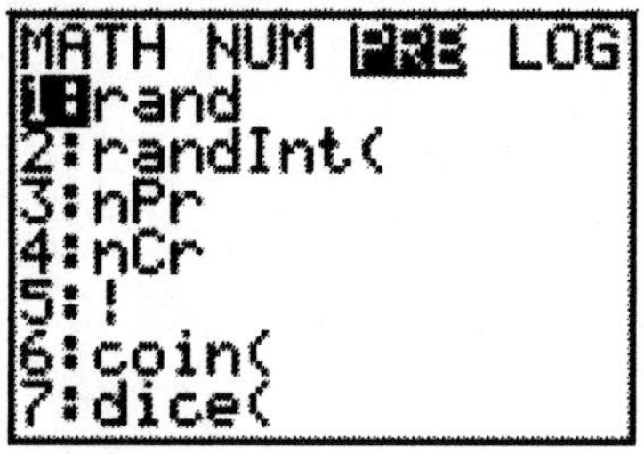

4. Select **2:randInt(** by pressing **2**.

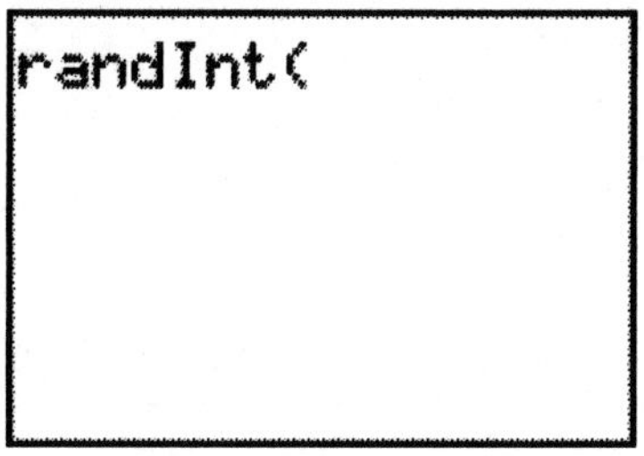

5. Press **1** [,] **4** [,] **20** [)] [STO▶] [2nd] [STAT] **2:L2**.

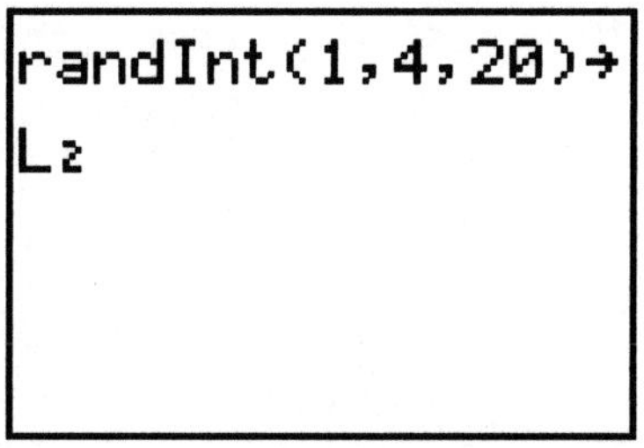

6. Press [ENTER] to generate the list of numbers and store them in **L2**.

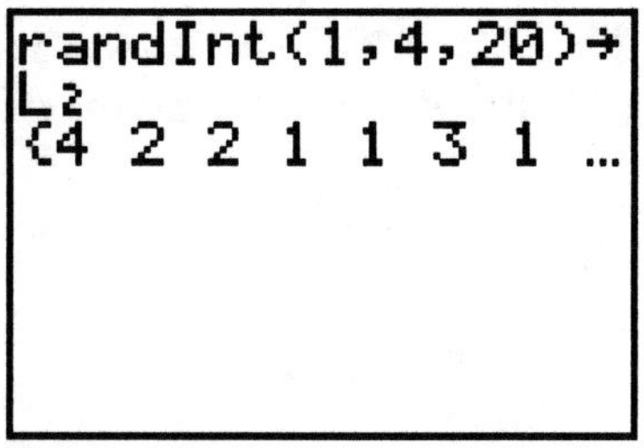

Compare your answers with the correct answers for the test. Using the equal sign, compare the number in **L1** to the corresponding number in **L2**. If the two values are *equal*, the TI-73 returns a *1*, which indicates that the statement is *true*. If the values are *not equal*, the TI-73 returns a *0*, which indicates that the statement is *false*. Use the following steps to perform this operation. Since this simulation will be repeated, you will save the formula that performs the operation.

7. Press [LIST]. Press [▶] [▶] [▲] to move the cursor to highlight **L3**.

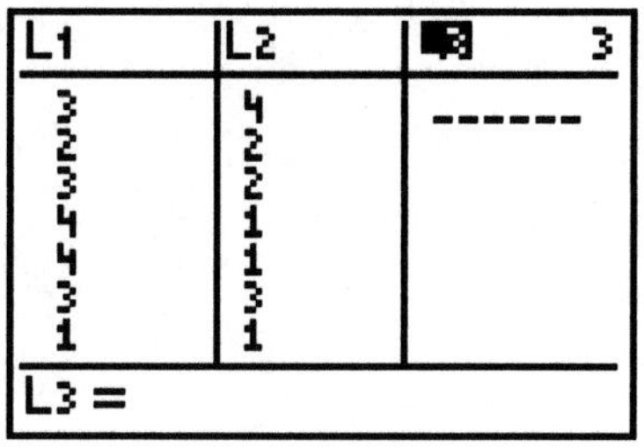

8. Press [2nd] [TEXT].

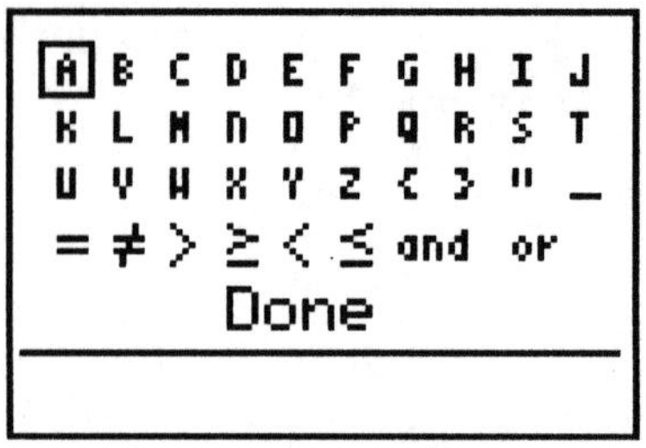

9. Press [◄] [◄] [▼] [▼] [ENTER] to select the quotation mark (").

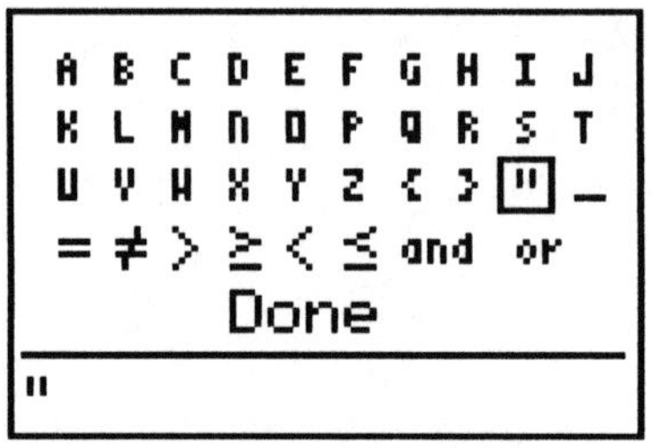

10. Press [▼] [▼] to move the cursor to **Done**.

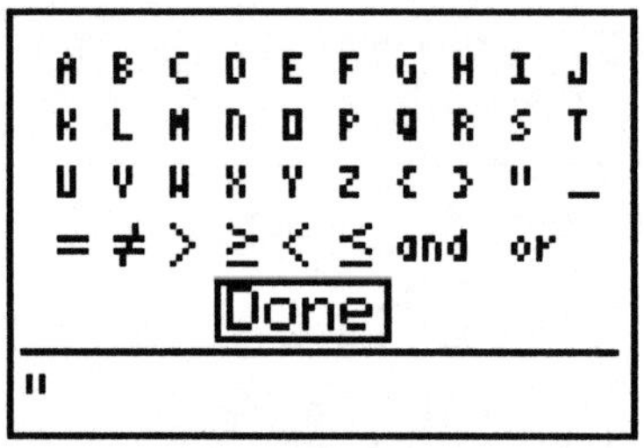

11. Press [ENTER] to exit the Text editor and paste the quotation mark in **L3**.

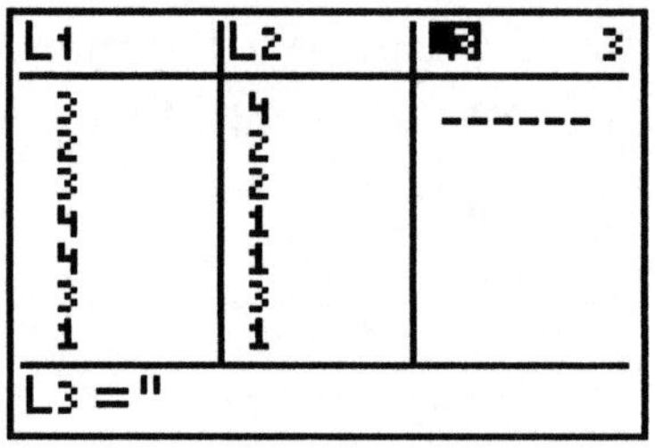

12. Press [2nd] [STAT]. Select **1:L1** by pressing **1** or [ENTER].

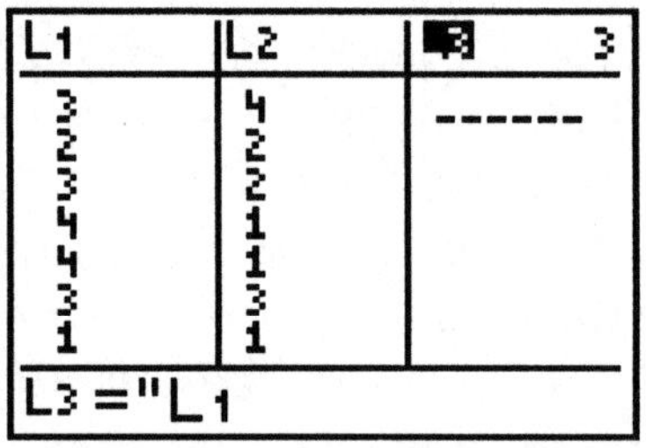

13. Press [2nd] [TEXT]. Press [▲] [▲] [ENTER] to select the equal sign (=).

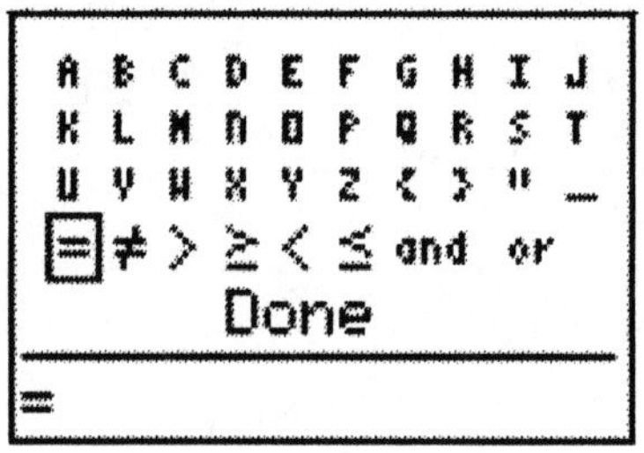

14. Press [▼] [ENTER] to exit the Text editor and paste the equal sign next to **L1**.

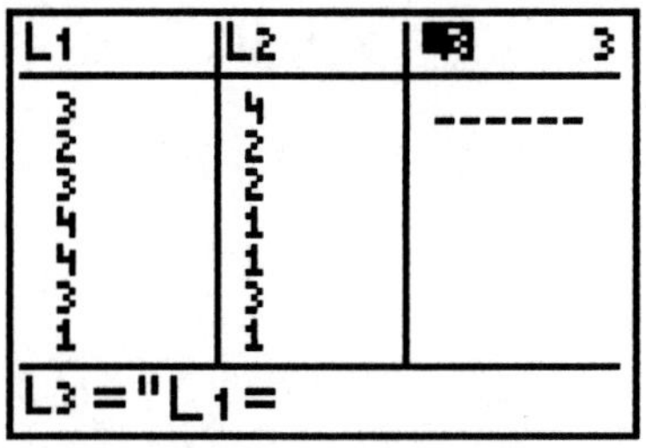

15. Press [2nd] [STAT]. Select **2:L2** by pressing **2**.

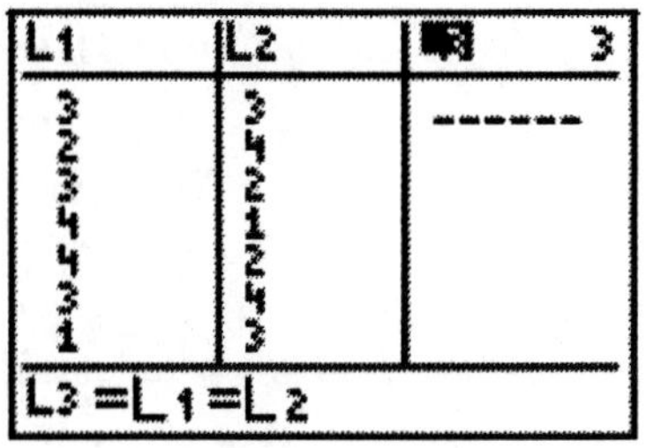

16. Press [2nd] [TEXT] [◄] [◄] [▼] [▼] [ENTER] to select the quotation mark ("). Press [▼] [▼] [ENTER] to exit the Text editor and paste the quotation mark next to **L2**.

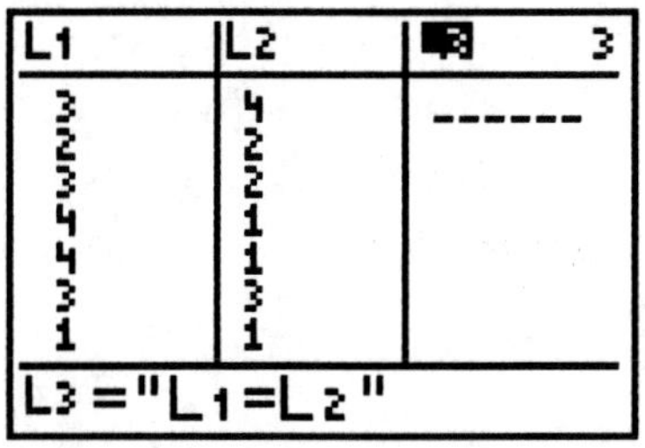

17. Press [ENTER] to see the comparison with the correct answers.

*Note: The symbol next to **L3** indicates a formula has been stored in **L3**.*

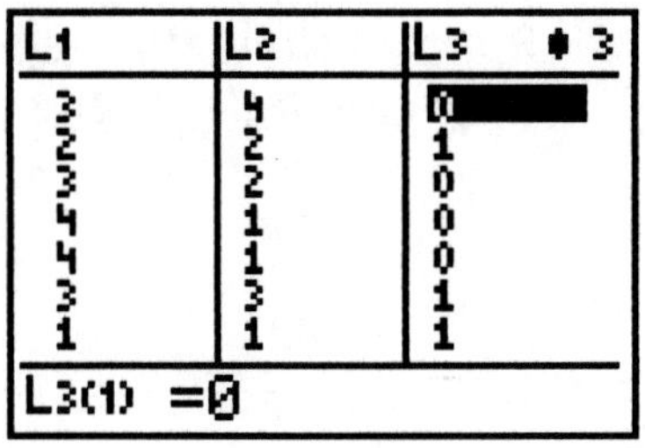

To find the number of correct answers in the simulation, calculate the sum of the numbers in **L3**.

18. Press [2nd] [QUIT] to exit the List editor.

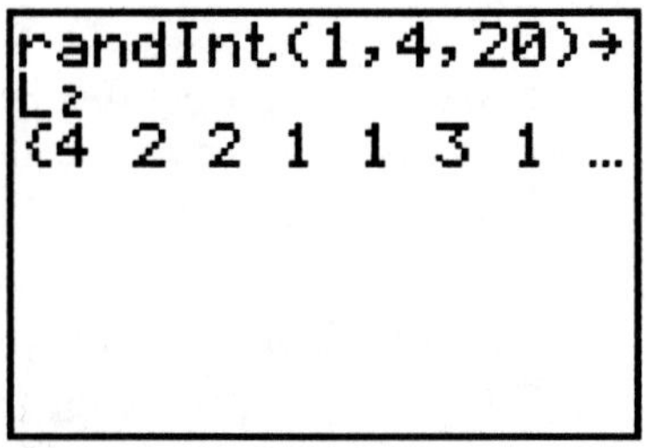

19. Press [2nd] [STAT] [►] [►] to move the cursor to the **MATH** menu.

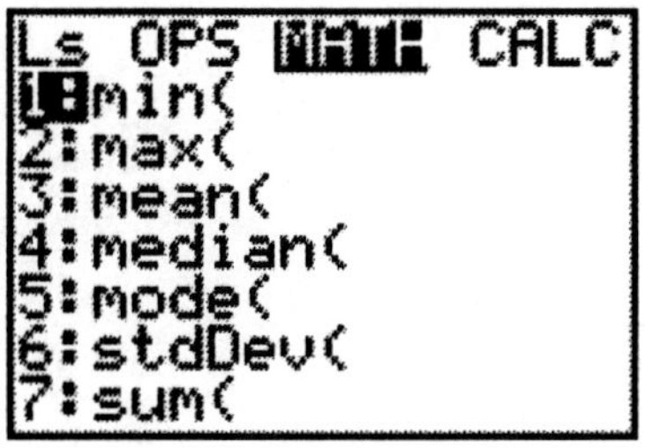

20. Select **7:sum(** by pressing **7**.

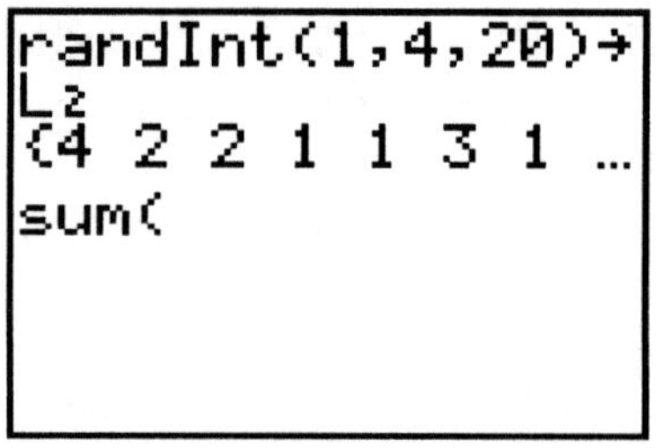

21. Press [2nd] [STAT] **3:L3** [)].

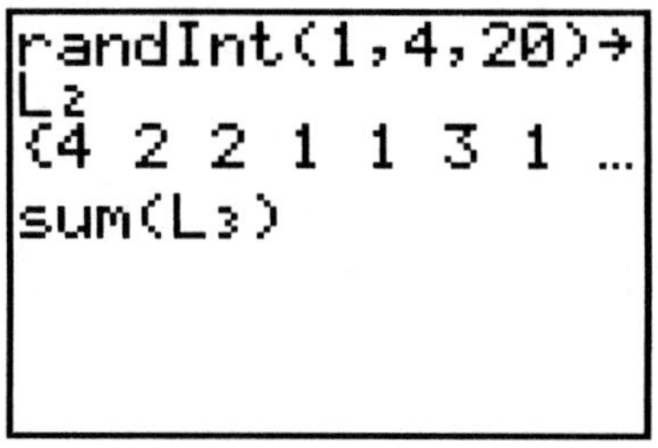

22. Press [ENTER] to see how many were correct. Record the data for this simulation in the table on the **Data Collection and Analysis** page.

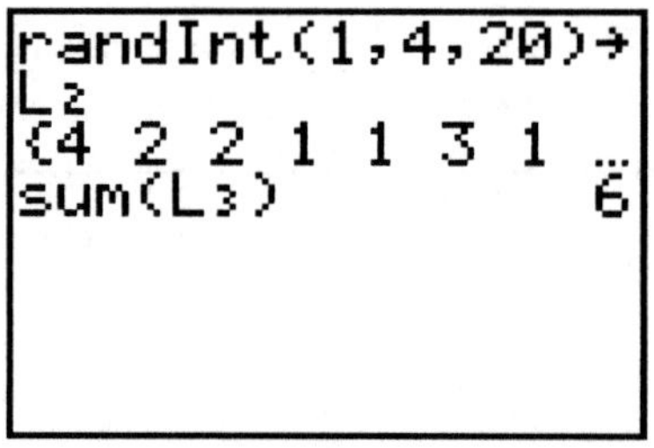

23. Run the simulation again. Press [▲] [▲] [▲] [▲] until **randInt(1,4,20)→L2** is highlighted.

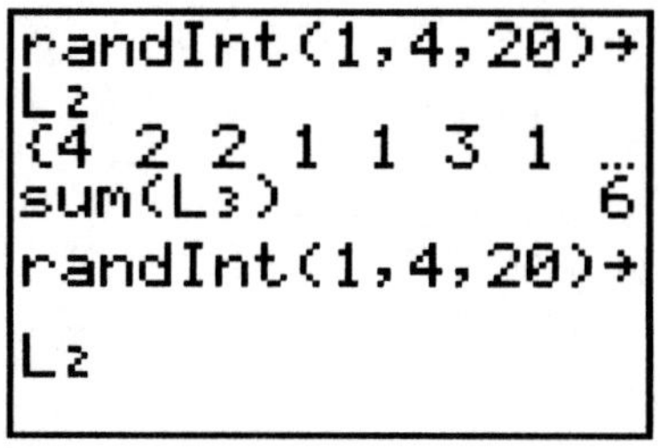

24. Press [ENTER] to copy the **randInt(** command, the press [ENTER] to run the simulation again.

Note: You can view the results of the new simulation by pressing [LIST]. When finished, press [2nd] [QUIT] to exit the List editor.

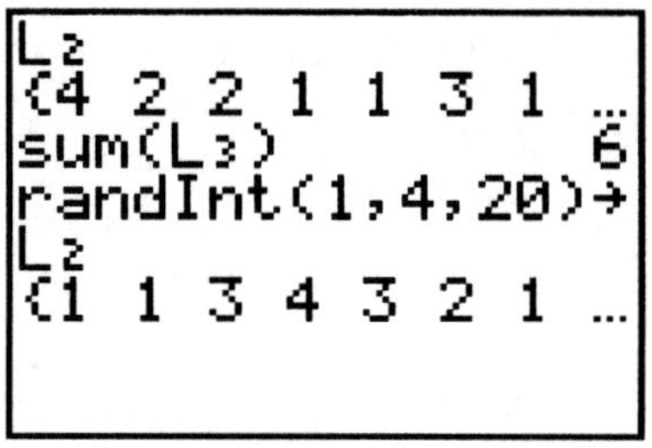

25. To calculate how many are correct, press [▲] [▲] [▲] [▲] until **sum(L3)** is highlighted.

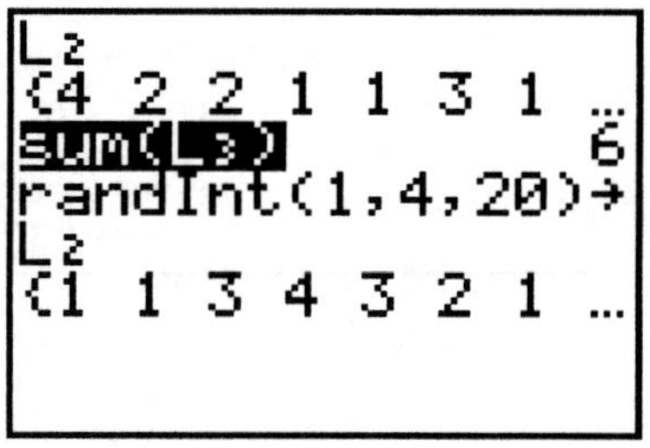

26. Press [ENTER] to copy the **sum(L3)** command, then press [ENTER] to calculate the number correct. Record the data for this simulation in the table on the **Data Collection and Analysis** page.

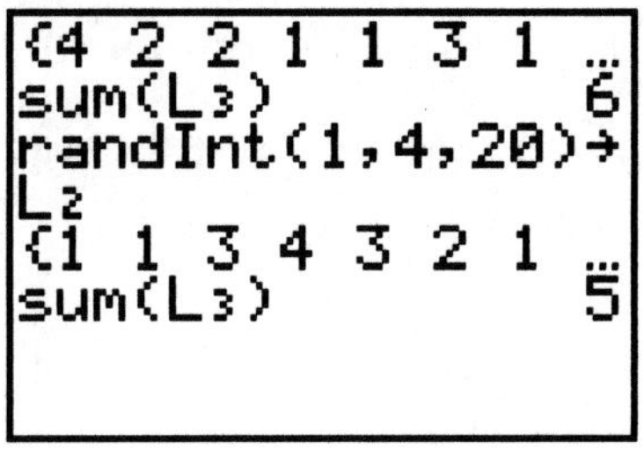

27. Run the simulation 40 to 50 more times (Steps 23-26). Record each of the trials on the **Data Collection and Analysis** page.

Setting up the window for the Histogram

1. Press [WINDOW] to set up the proper scale for the axes.

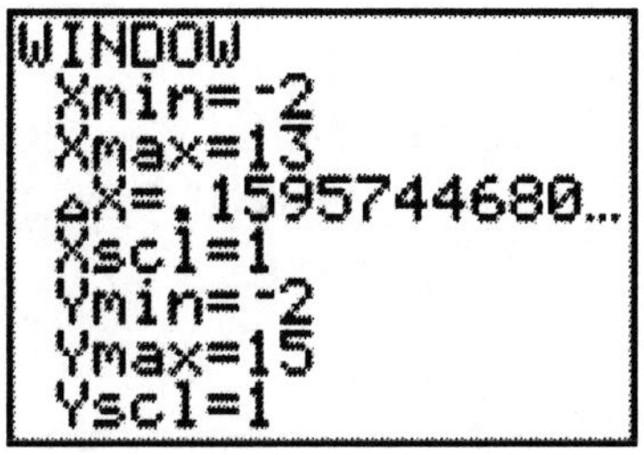

2. Set the **Xmin** value by identifying the minimum number of correct answers from the **Data Collection and Analysis** page. Choose a number that is less than the minimum.
3. Set the **Xmax** value by identifying the maximum number of correct answers from the **Data Collection and Analysis** page. Choose a number that is greater than the maximum. **Do Not Change the ΔX Value.** Set the **Xscl** to **1**.
4. Set the **Ymin** value by identifying the minimum in the frequency column from the **Data Collection and Analysis** page. Choose a number that is less than the minimum.
5. Set the **Ymax** value by identifying the maximum value in the frequency column from the **Data Collection and Analysis** page. Choose a number that is greater than the maximum. Set the **Yscl** to **1**.

Graphing the data: Plotting a histogram

Use the data in the table on the **Data Collection and Analysis** page, Part I, to plot a histogram.

1. Press [LIST].

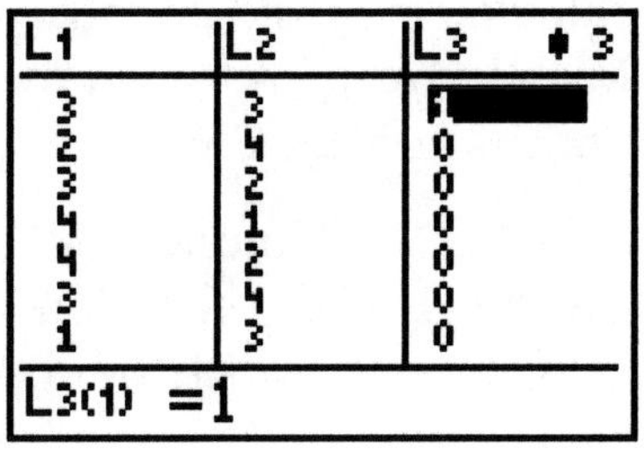

2. Enter the number of correct answers in **L4**.
3. Enter the frequencies in **L5**.

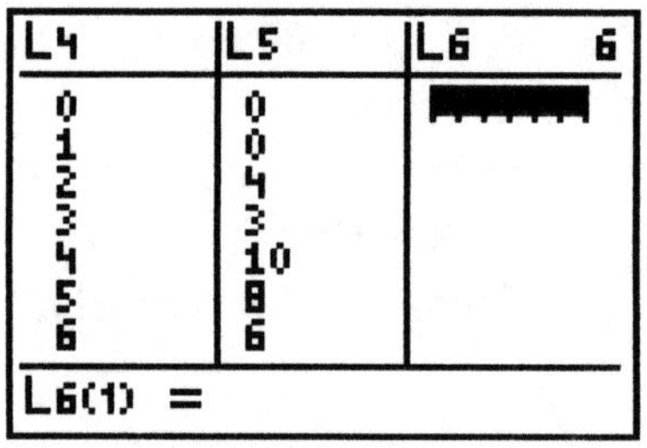

4. Press [2nd] [PLOT]. Select **1:Plot1** by pressing **1** or [ENTER].

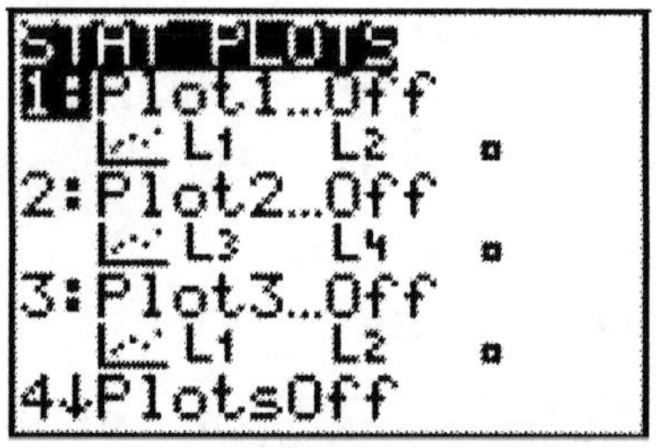

5. Set up the plot as shown by pressing [ENTER] [▼] [▶] [▶] [▶] [▶] [▶] [ENTER] [▼] [2nd] [STAT] **4:L4** [▼] [2nd] [STAT] **5:L5.**

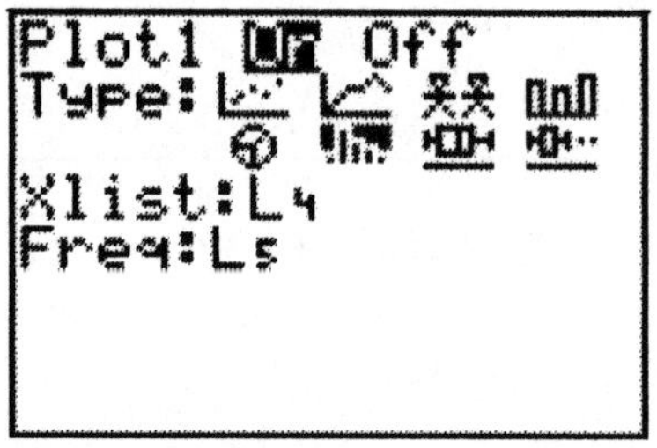

6. Press [TRACE] to see the plot. Use [◀] and [▶] to see the frequencies.

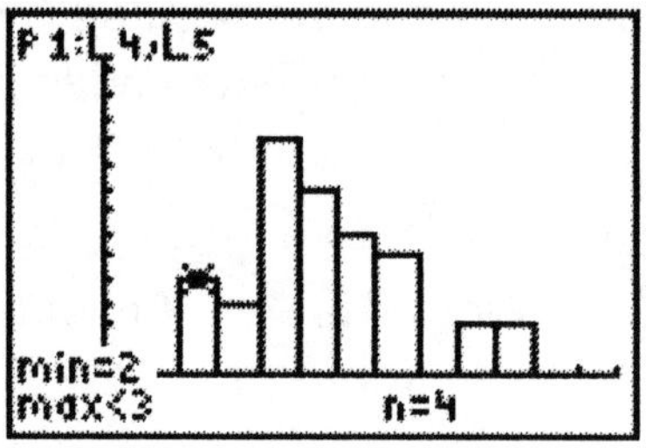

Answer Part I questions 1 – 9 on the **Data Collection and Analysis** page.

Collecting the data – Part II

Nathan noticed that the correct answers on the test contained more C's and D's than A's and B's. He decided to run a simulation on his calculator to see if the calculator would produce a distribution of A – D's similar to that on the test. Nathan decided to run three practice tests for his simulation and look at the number of A's, B's, C's, and D's.

Use the TI-73 and the random number generator to simulate the answers on three practice tests. You will plot a pie graph, a pictograph, and a bar graph to determine the distribution of letters on the test.

Setting up the TI-73

Before starting your data collection, make sure that the TI-73 has the STAT PLOTS turned OFF, Y= functions turned OFF or cleared, the MODE and FORMAT set to their defaults, and the lists cleared. See the Appendix for a detailed description of the general setup steps.

1. Press [MATH].

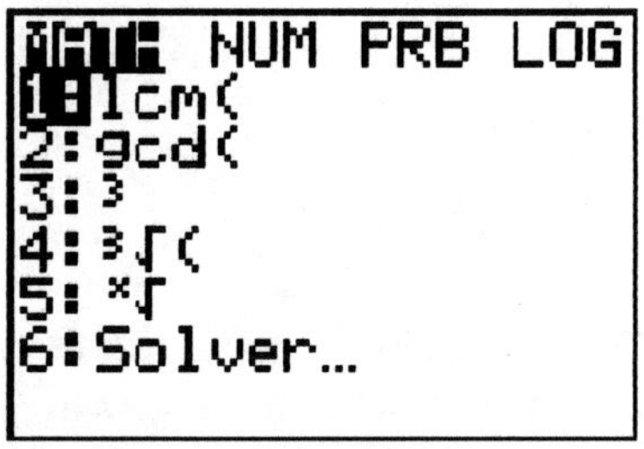

2. Press [▶] [▶] to move the cursor to the **PRB** menu.

3. Select **2:randInt(** by pressing **2**.

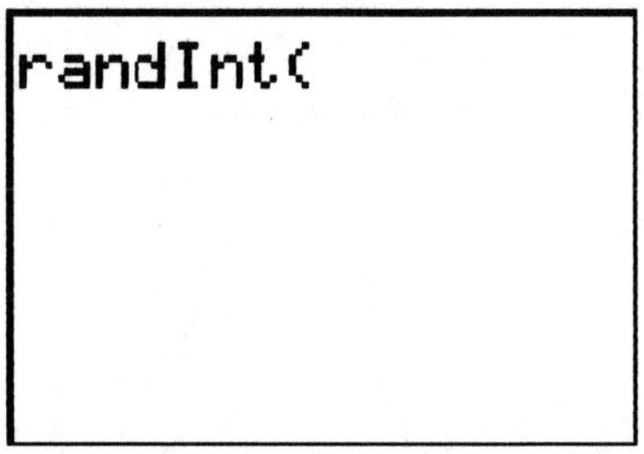

4. Press **1** [,] **4** [,] **20** [)] [STO▶] [2nd] [STAT] **1:L_1**. Press [ENTER] to generate a list of numbers that represent the answers to a practice test 1.

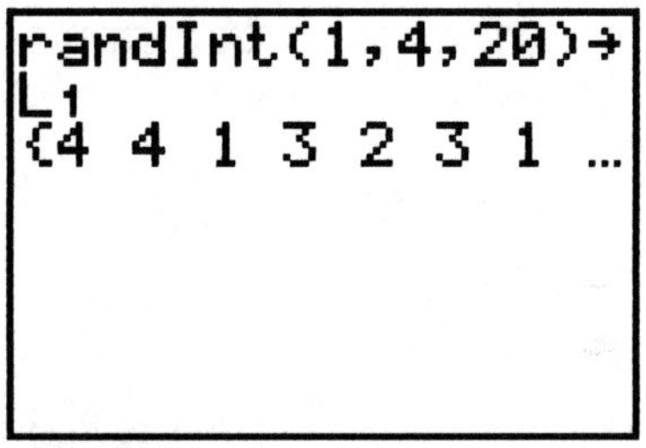

5. Sort the list. Press [2nd] [STAT] [▶] to move the cursor to the **OPS** menu.

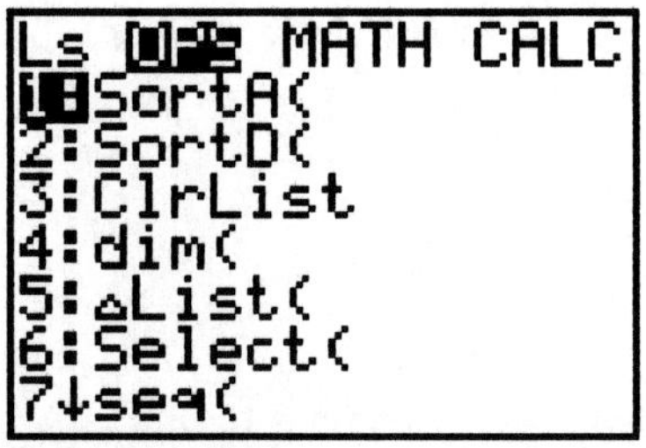

6. Select **1:SortA(** by pressing **1** or [ENTER].

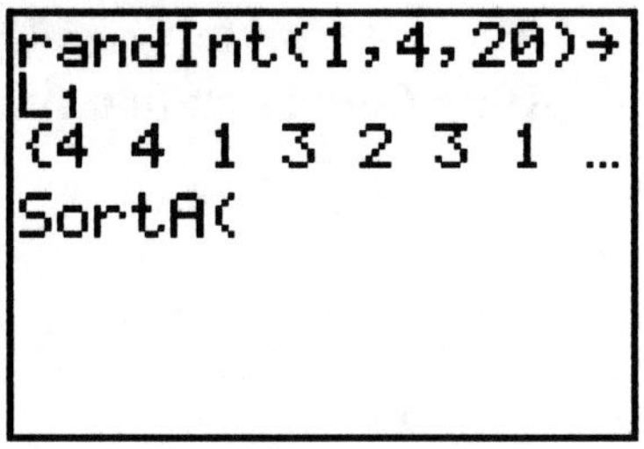

7. Press [2nd] [STAT] **1:L1** [)] [ENTER].

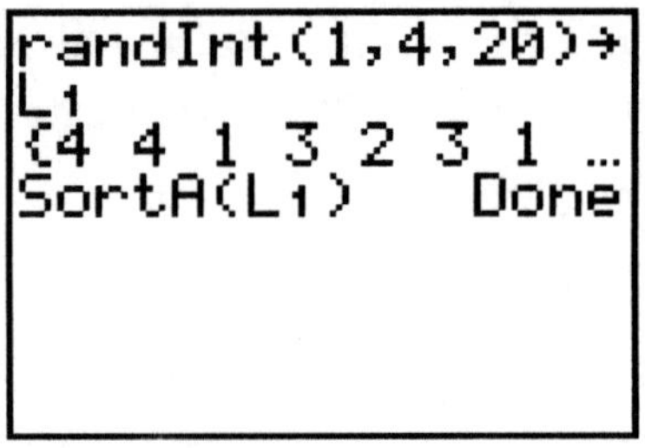

8. Press [LIST]. Count the number of A's, B's, C's, and D's. (Remember that A = 1, B = 2, C = 3, and D = 4.)

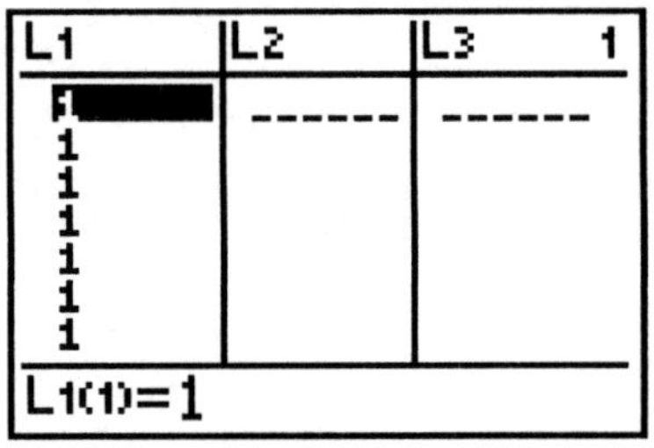

Enter the data in the table for Part II on the **Data Collection and Analysis** page.

9. Press [2nd] [QUIT] to return to the Home screen. Press [2nd] [ENTRY] twice until you get the **randInt(1,4,20)→L1** statement on the screen.

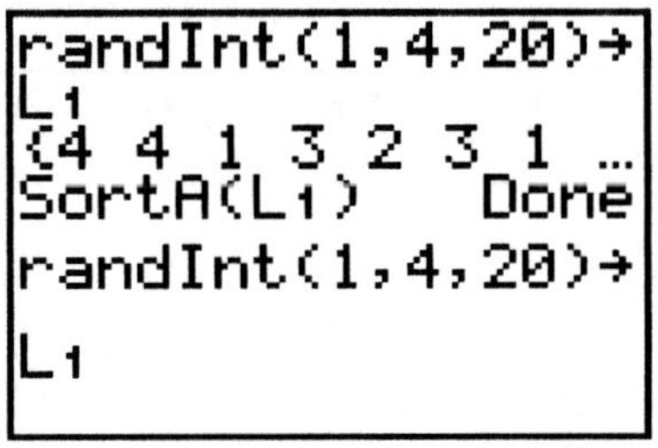

10. Press [ENTER] to generate a second list of numbers that represent the answers to a practice test 2.

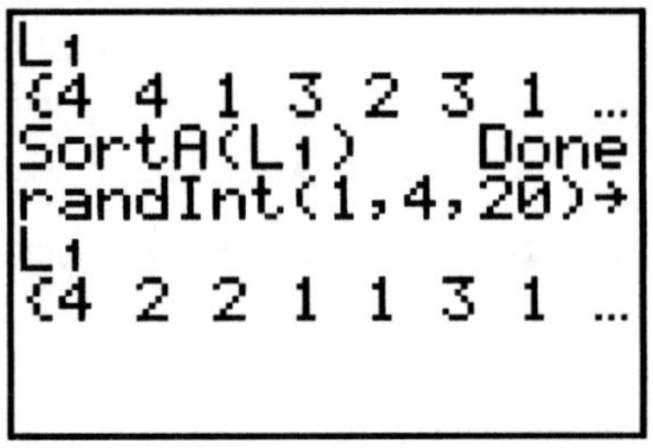

11. Press [2nd] [ENTRY] twice until you get the **SortA(L1)** statement.

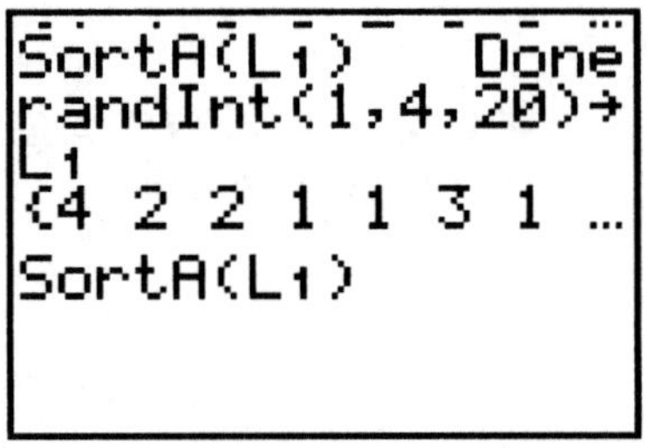

12. Press [ENTER] to sort the data. Repeat Step 8 and record your results in the table on the **Data Collection and Analysis** page.

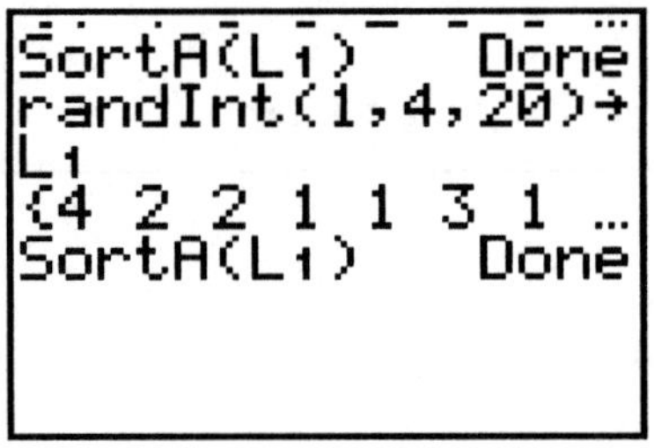

13. Repeat Steps 9 - 12 to generate the answers to a practice test 3. Record your results in the table on the **Data Collection and Analysis** page.

Entering the data in the TI-73

1. Press [LIST] and press [◄] to place the cursor at the top of the 7th list.

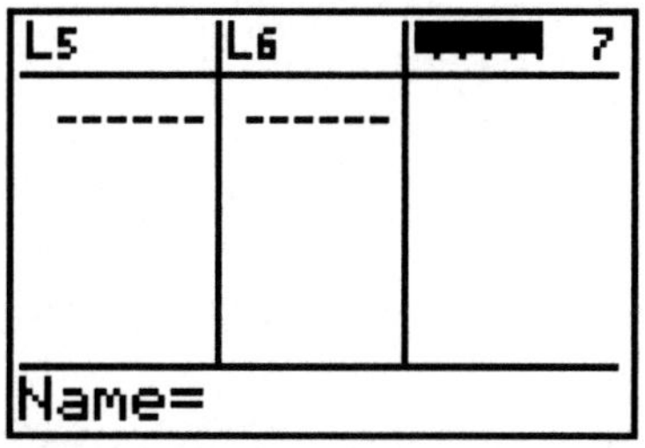

2. Name the list **ANSWR** by pressing [2nd] [TEXT], moving the cursor to each letter of the name **A N S W R,** and pressing [ENTER].

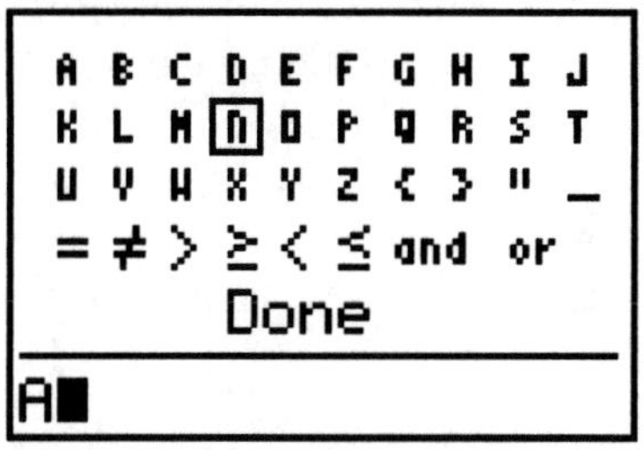

3. Move the cursor to highlight **DONE**.

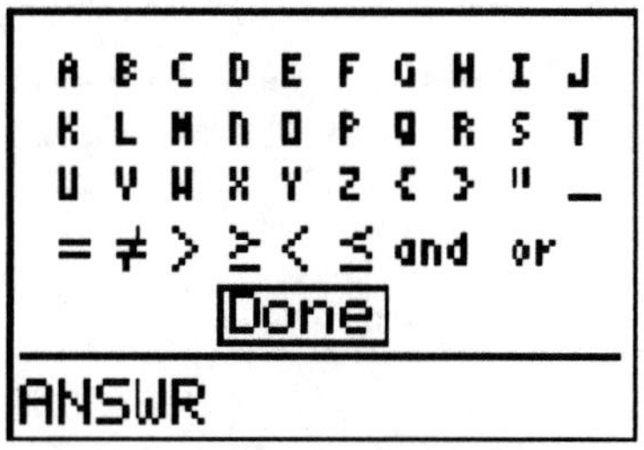

4. Press [ENTER] to exit the Text editor.

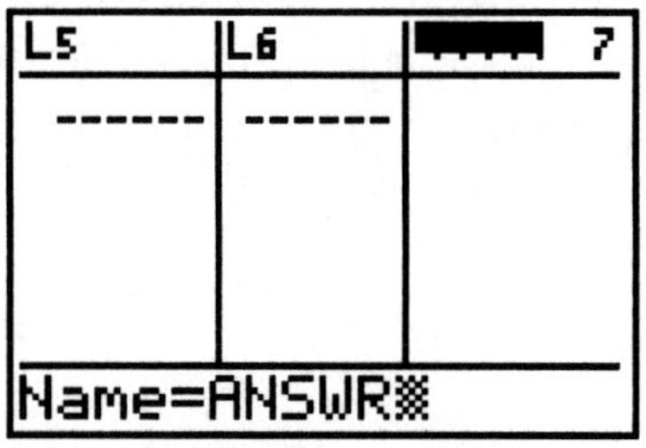

5. Press [ENTER] to paste **ANSWR** at the top of the list.

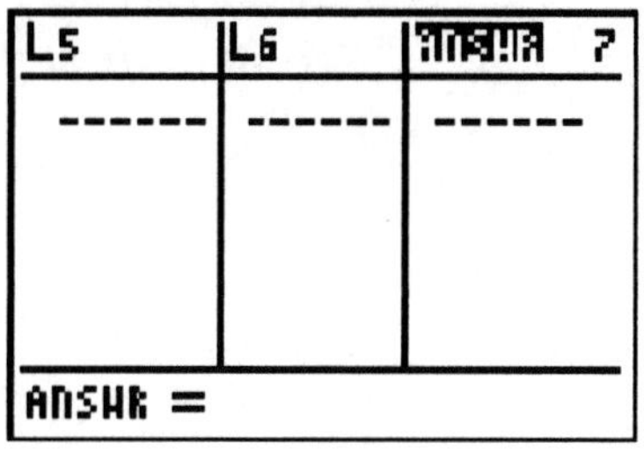

Create a category list (a list that contains text) by having the first element entered in quotation marks.

6. Press [▼] to move the cursor to the first element.

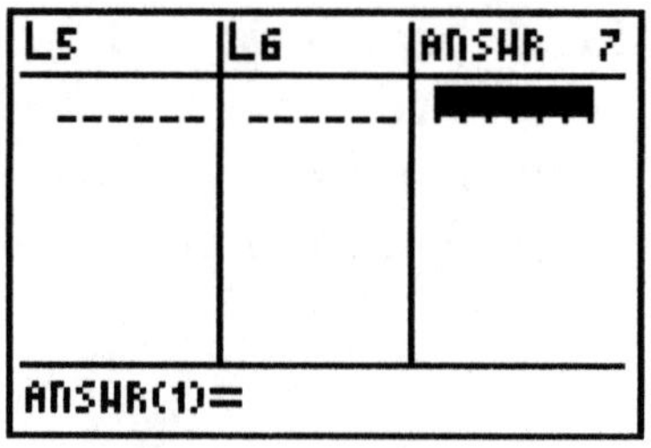

7. Enter **"A"**. Press [2nd] [TEXT]. Move the cursor to each character of the entry **"A"** and press [ENTER].

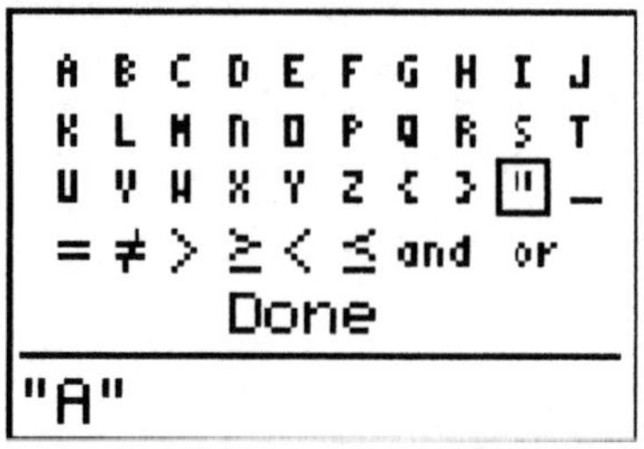

8. Move the cursor to highlight **DONE.** Press [ENTER] to exit the Text editor.

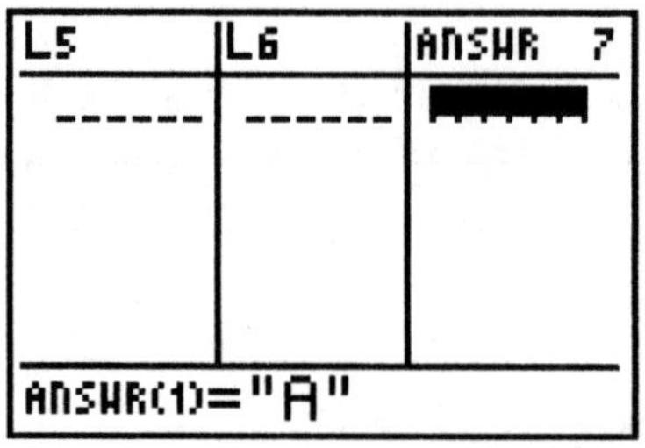

9. Press [ENTER] to paste **A** in the list.

Note: *A* ***c*** *should appear at the top of the list indicating that this is a category list.*

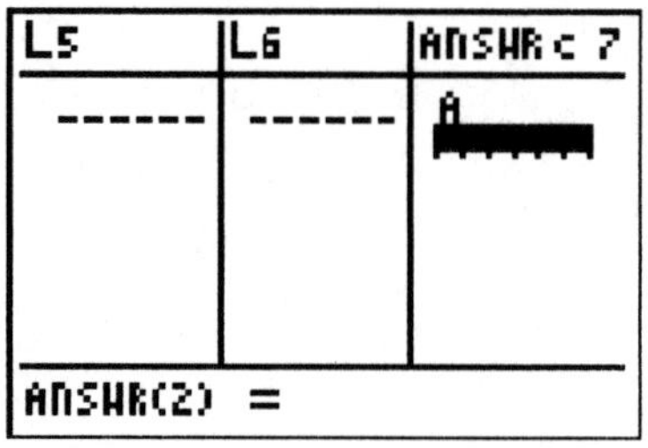

10. Enter the letters **B**, **C**, and **D** in the list by repeating Steps 7-9. You DO NOT have to enclose the remaining letters in quotation marks.

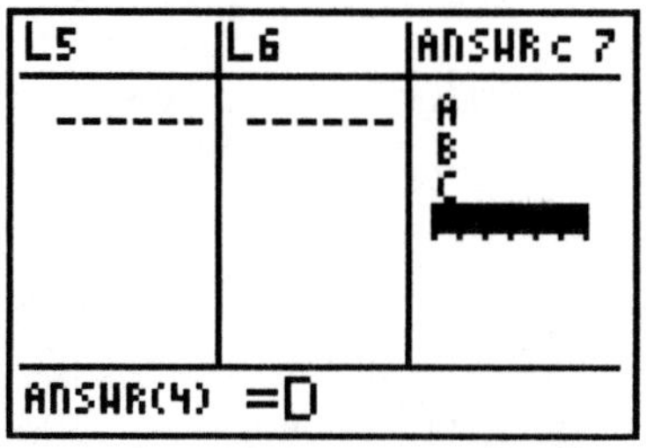

11. Press [▶] to move the cursor to the top of the 8th list. Repeat Steps 1-5 using the list name,**TEST1**.

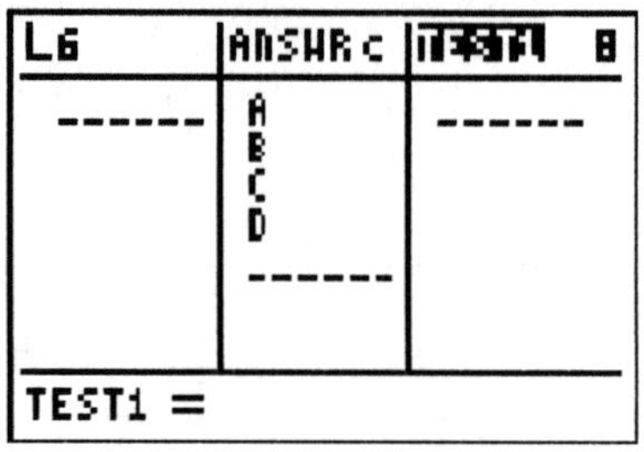

12. Repeat Step 11 for the list names **TEST2** (9th list) and **TEST3** (10th list.)

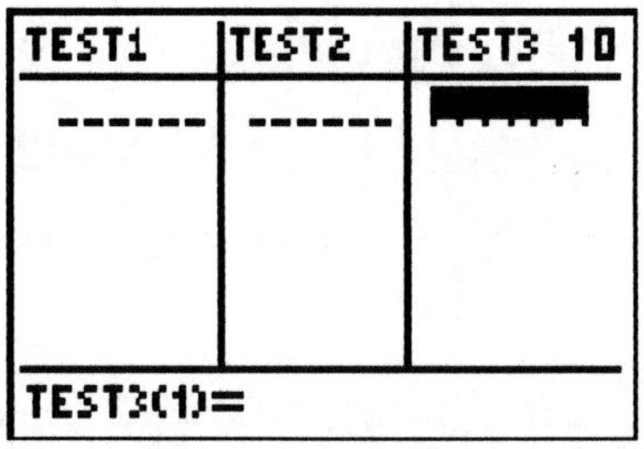

13. Enter the data from the table in Part II of the **Data Collection and Analysis** page in the appropriate lists.

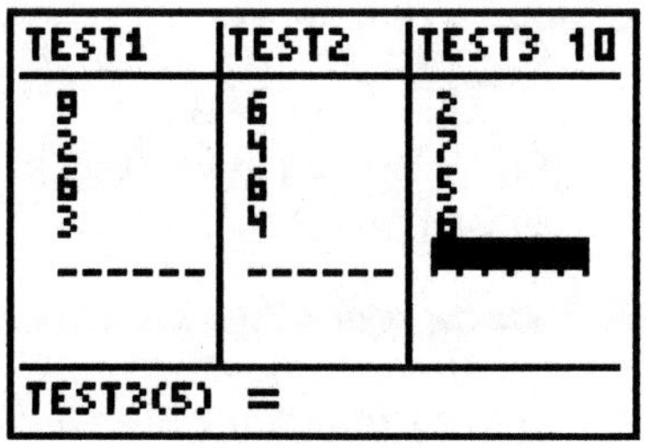

Graphing the data: Setting up a pie chart

Use the data in the table on the **Data Collection and Analysis** page Part II to plot a pie chart.

1. To set up the plot, press [2nd] [PLOT]. Select **Plot1** by pressing **1** or [ENTER].

Press [ENTER] [▼] [▶] [▶] [▶] [▶] [ENTER] [▼] [2nd] [STAT] **7:ANSWR** [▼] [2nd] [STAT] **8:TEST1** [▼] [▶] [ENTER].

Note: *Your lists, ANSWR and TEST1 may not be in positions 7 and 8 on the TI-73. Use [▼] and [▲] to move the cursor to the desired list and press [ENTER] to select that list.*

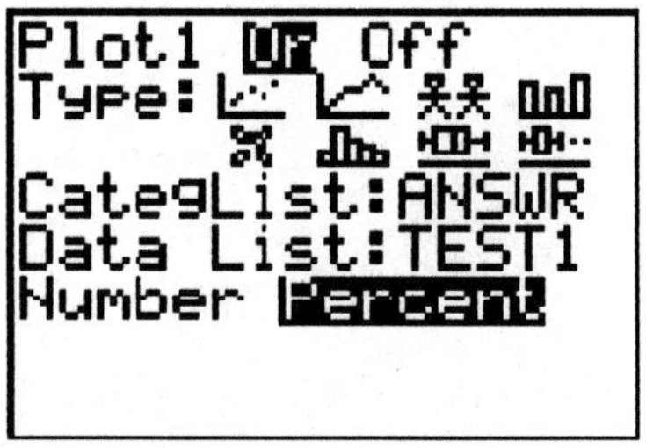

2. Press [TRACE] to see the pie chart.

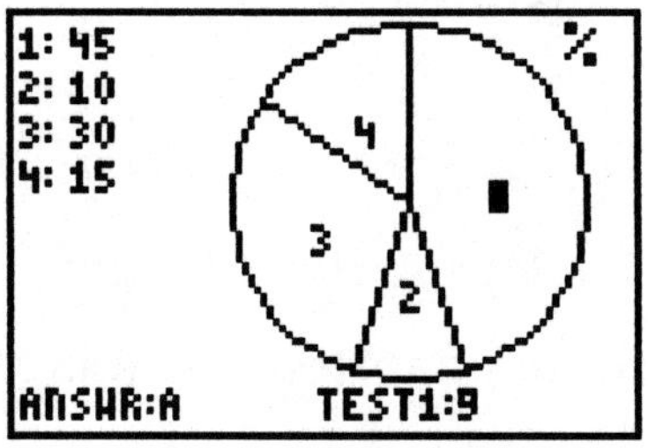

3. Use [◀] and [▶] to see the number of items in each section of the graph. The numbers displayed in the left hand corner represent the percent (%) for each letter.

4. To view the pie chart for the data in **TEST2,** press [2nd] [PLOT]. Select **Plot1** by pressing **1** or [ENTER]. Press [▼] [▼] [▼] [▶] [▶] [▶] [▼] **2.**

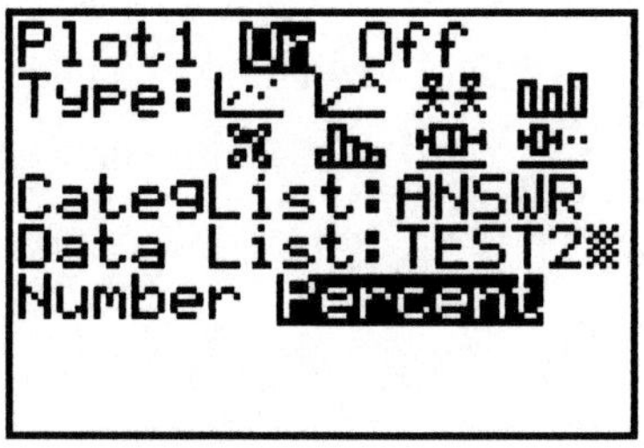

5. Repeat Steps 2-3 using the data from **TEST2.**

6. Repeat Steps 4-5 using the data from **TEST3.**

Answer Part II question 1 on the **Data Collection and Analysis** page.

Graphing the data: Setting up a pictograph

1. Press [2nd] [PLOT]. Select **Plot1** by pressing **1** or [ENTER]. Press [ENTER] [▼] [▶] [▶] [ENTER] [▼] [2nd] [STAT] **7:ANSWR** [▼] [2nd] [STAT] **8:TEST1** [▼] 2 [▼] [▶] [ENTER] [▼] [ENTER].

   Note: *Your lists,* ***ANSWR*** *and* ***TEST1,*** *may not be in positions 7 and 8 on the TI-73. Use [▼] and [▲] to move the cursor to the desired list and press [ENTER] to select that list.*

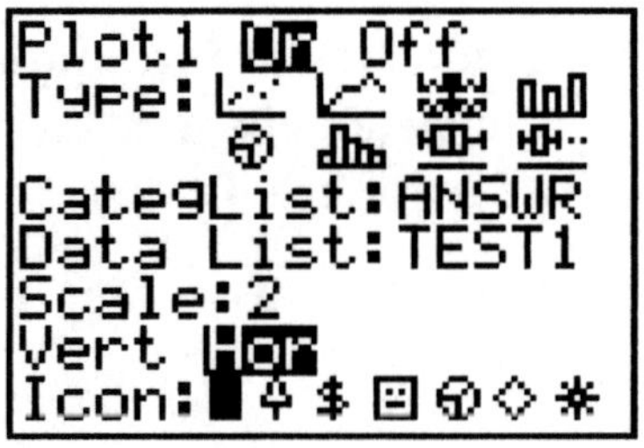

2. Press [TRACE] [▶] to see the pictograph.

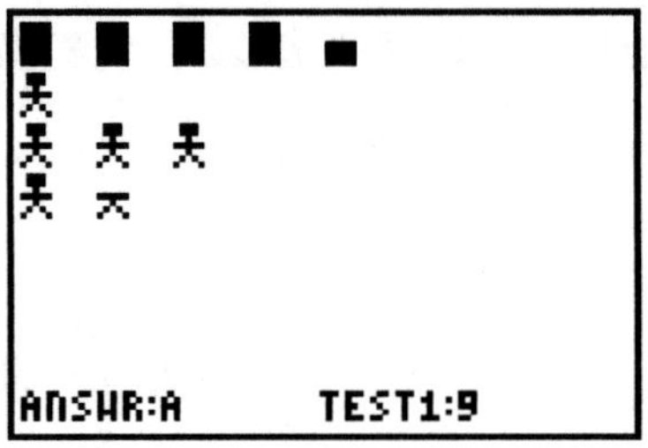

3. Use [▶] and [◀] to see the number of items represented by each row of data.

4. To view the pictographs for the data in **TEST2,** press [2nd] [PLOT]. Select **Plot1** by pressing **1** or [ENTER]. Press [▼] [▼] [▼] [▶] [▶] [▶] [▼] **2.**

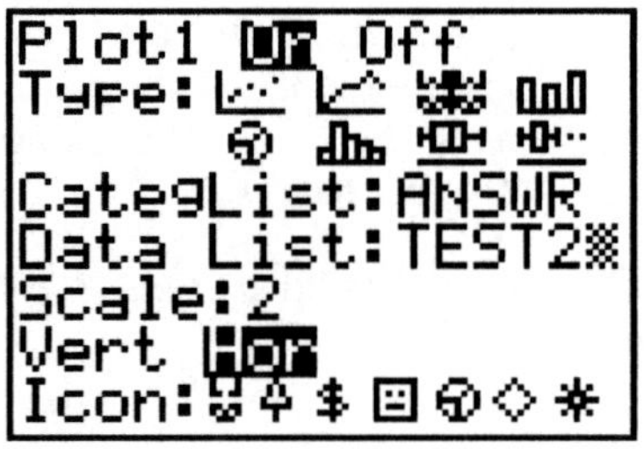

5. Repeat Steps 2-3 using the data from **TEST2.**

6. Repeat Steps 4-5 using the data from **TEST3.**

Answer Pat II question 2 on the **Data Collection and Analysis** page.

Graphing the data: Setting up a bar graph

1. To set up the plot, press [2nd] [PLOT]. Select **Plot1** by pressing **1** or [ENTER]. Press [ENTER] [▼] [▶] [▶] [▶] [ENTER] [▼] [2nd] [STAT] **7:ANSWR** [▼] [2nd] [STAT] **8:TEST1** [▼] **9:TEST2** [▼] **0:TEST3** [▼] [ENTER] [▶] [▶] [▶] [▶] [ENTER].

   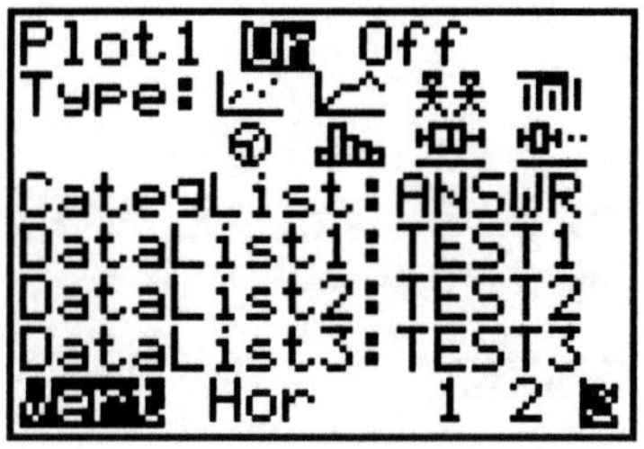

   ***Note**: Your lists, **ANSWR**, **TEST1**, **TEST2**, and **TEST3**, may not be in positions 7, 8, 9, and 0 (10) on the TI-73. Use [▼] and [▲] to move the cursor to the desired list and press [ENTER] to select that list.*

2. Press [TRACE] to see the bar graph.

   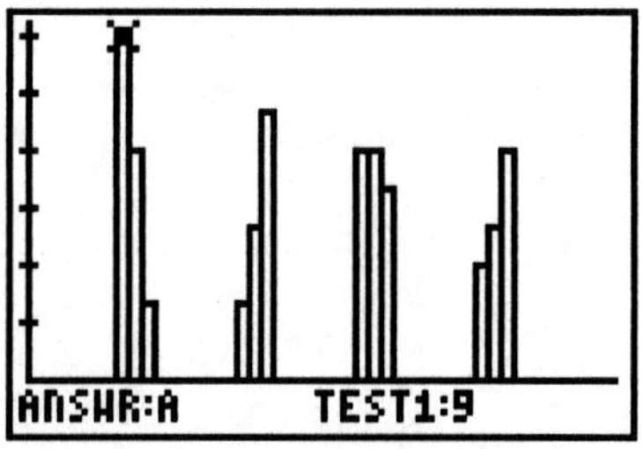

3. Use [▶] and [◀] to see the number of items represented by each bar of data.

Answer Part II questions 3 and 4 on the **Data Collection and Analysis** page.

Data Collection and Analysis

Name ____________________

Date ____________________

Activity 9: You're Probably Right, It's Wrong

Collecting the data – Part I

Record your data in the table below.

Number of correct answers	Tally marks	Frequency
0		
1		
2		
3		
4		
5		
6		
7		
8		
9		
10		
11		
12		
13		

Analyzing the data – Part I

1. Find the *mean* for the number of correct answers. ____________________

   (Press [2nd] [QUIT] [CLEAR] [2nd] [STAT] [▸] [▸] **3:mean(** [2nd] [STAT] **4:L4** [,] [2nd] [STAT] **5:L5** [)] [ENTER])

2. Find the *median* for the number of correct answers. ____________________

   (Press [2nd] [STAT] [▸] [▸] **4:median(** [2nd] [STAT] **4:L4** [,] [2nd] [STAT] **5:L5** [)] [ENTER])

3. Find the *mode* for the number of correct answers. ____________________

   (Press [2nd] [STAT] [▸] [▸] **5:mode(** [2nd] [STAT] **4:L4** [,] [2nd] [STAT] **5:L5** [)] [ENTER])

4. Which measure of central tendency do you think gives a better indication of what might happen if you use this method to answer the questions on a multiple-choice test? Explain your answer.

____

____

5. Using your answer from number 4, find the *experimental probability*.

____

6. Find the *theoretical probability*. _______________________

7. How do your answers in number 5 and number 6 compare?

____

____

8. Do you think it is a good idea to use a random number generator to answer the multiple-choice questions on a test? Explain.

____

____

9. Write a random number statement to answer 20 True / False questions.

randInt(_______________________________________**)**

Collecting the data – Part II

Answer	Test 1 Amount	Test 2 Amount	Test 3 Amount
A			
B			
C			
D			

Analyzing the data – Part II

1. Compare the percentage of A's from the pie charts. Are the percentages the same or different for each test? Would you expect them to be the same or different? Explain.

2. Compare the number of B's for each practice test using the pictograph. Are the number of B's the same or different for each test? Do you think the pictograph is a good way of comparing the data? Explain.

3. Compare the number of D's for each practice test using the bar graph. Are the number of D's the same or different for each test? Explain.

4. Which graph, pie chart, pictograph, or bar graph is best for comparing the number of A's, B's, C's, and D's in each test? Explain.

Teacher Notes

Activity 9

You're Probably Right, It's Wrong

Objectives

- To use technology to find experimental and theoretical probabilities
- To use technology to find measures of central tendencies
- To use technology to explore simulation
- To use technology to generate random numbers
- To use technology to plot a histogram
- To use technology to plot a pie chart
- To use technology to plot a pictograph
- To use technology to plot a bar graph

Materials

- TI-73 graphing device

Preparation – Part I

- Make sure students run enough trials to produce at least 40 to 50 data items.
- Find the *mean, median, and mode* by using the [2nd] [STAT] [MATH] menu on the TI-73. Check students' results for finding the mean.
- For a histogram, discuss the values at the bottom of the screen for the plot (that is, the values of *n*, *min*, and *max*).

Preparation – Part II

- After the activity, to remove the list that has an equation stored in it, press [2nd] [STAT] [▸] **3:ClrList** [2nd] [STAT] **3:L3** [ENTER].

Answers to Data Collection and Analysis questions

Collecting the data – Part I

Sample data:

Number of correct answers	Tally marks	Frequency
0	-	0
1	-	0
2	////	4
3	///	3
4	~~////~~ ~~////~~	10
5	~~////~~ ///	8
6	~~////~~ /	6
7	~~////~~	5
8	-	0
9	//	2
10	//	2
11		
12		
13		

Analyzing the data – Part I

1. Find the *mean* for the number of correct answers.

   Per the sample data, mean = 5.15

2. Find the *median* for the number of correct answers.

   Per the sample data, median = 5

3. Find the *mode* for the number of correct answers.

   Per the sample data, mode = 4

4. Which measure of central tendency do you think gives a better indication of what might happen if you use this method to answer the questions on a multiple-choice test? Explain your answer.

   The median or mean gives a better indication of the results of using this method to answer questions on a multiple-choice test. Answers may vary.

5. Using your answer from number 4, find the *experimental probability.*

   5 / 20 or .25

6. Find the *theoretical probability.*

   5 / 20 or .25

7. How do your answers in number 5 and number 6 compare?

They are the same. Answers may vary.

8. Do you think it is a good idea to use a random number generator to answer multiple-choice questions on a test? Explain.

No. The TI-73 simulates the theoretical probability, which means that the score will most likely be around 25%.

9. Write a random number statement to answer 20 True/ False questions.

randInt(*1,2,20*)

Collecting the data – Part II

Answer	Test 1 Amount	Test 2 Amount	Test 3 Amount
A	9	6	2
B	2	4	7
C	6	6	5
D	3	4	6

Analyzing the data – Part II

1. Compare the percentage of A's from the pie charts. Are the percentages the same or different for each test? Would you expect them to be the same or different? Explain.

The percentages are different. You would expect the percentages to be approximately the same. Since each answer has an equally likely chance of occurring, you would expect 25% of the answers to be A for each trial.

2. Compare the number of B's for each practice test using the pictograph. Are the number of B's the same or different for each test? Do you think the pictograph is a good way of comparing the data? Explain.

The number of B's is different. Answers will vary.

3. Compare the number of D's for each practice test using the bar graph. Are the number of D's the same or different for each test? Explain.

The number of D's is different. Answers will vary.

4. Which graph on the TI-73, pie chart, pictograph, or bar graph, is best for comparing the number of A's, B's, C's, and D's in each test? Explain.

The bar graph is the best graph for comparison because it is the only one on the TI-73 that allows you to clearly see the results of all three tests, side-by-side.

EXPLORATIONS

Activity 10

That's a Stretch

Objectives

- To determine the relationship between the stretch of a spring and the number of weights in a cup suspended from the spring
- To find the *y* value of a function, given the *x* value
- To find the *x* value of a function, given the *y* value
- To use technology to find a best fit line
- To use technology to plot a set of ordered pairs

Materials

- TI-73 graphing device
- Slinky® cut in half, one per group
- Small bathroom paper cups or film containers, one per group
- Marbles, pennies, or other small objects such as cubes, at least 40 per group
- Large paper clips, one per group
- Meter stick, one per group

Introduction

When you bounce a basketball, the shape of the ball temporarily changes. When you pluck a string on a guitar, the shape of the string changes. When a weight is suspended from a spring, the spring stretches. If additional weights are added, the spring stretches even more. Once the weights are removed, the spring returns to its original shape.

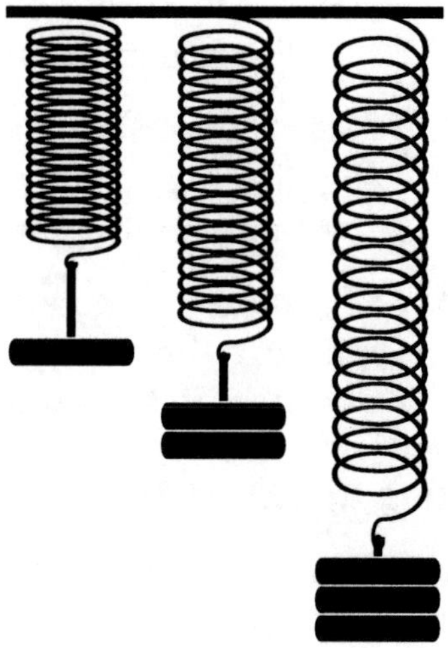

The basketball, the guitar string, and the spring are said to be *elastic*. If an external force is applied to an object, it creates stress within the object that causes it to become deformed. *Elasticity* is the property of a body that causes it to return to its initial size and shape after being compressed or stretched. Not all materials return to their initial state after a force is applied. These materials are said to be *inelastic*. Some examples of objects that are inelastic are clay, lead, and dough.

For many materials, the amount of stretch or compression is directly proportional to the applied force. This relationship was first expressed by British physicist Robert Hooke and is known as *Hooke's Law*.

Problem

Steel is an *elastic* material. Many springs are constructed of steel. What would happen if you suspended objects from a steel spring? Would the spring stretch at a constant rate or an exponential rate?

Collecting the data — Part I

Each group of students should obtain one meter stick, a cup, a Slinky®, and 40 marbles or other small objects from your teacher.

Using a paper clip, create a handle on the cup. Hang the cup on the Slinky. Place the Slinky on the center of the meter stick. Place the meter stick across two chairs as shown in the diagram below.

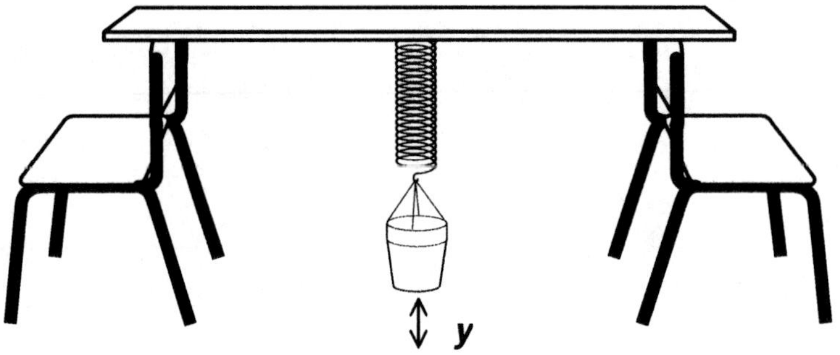

1. Measure the distance, in centimeters, from the floor to the bottom of the cup. Record the distance in the table on the **Data Collection and Analysis** page.
2. Place five of the objects that you are using in the cup. When the Slinky is stable, measure the distance from the floor to the bottom of the cup. Record the distance in the table on the **Data Collection and Analysis** page.
3. Place five additional objects in the cup. When the Slinky is stable, measure the distance from the floor to the bottom of the cup. Record the distance in the table on the **Data Collection and Analysis** page.
4. Continue placing additional objects in the cup in increments of five and measure the distance from the floor to the bottom of the cup. Record the distances in the table on the **Data Collection and Analysis** page.

Setting up the TI-73

Before starting your data collection, make sure that the TI-73 has the STAT PLOTS turned OFF, Y= functions turned OFF or cleared, the MODE and FORMAT set to their defaults, and the lists cleared. See the Appendix for a detailed description of the general setup steps.

Entering the data in the TI-73

1. Press [LIST].

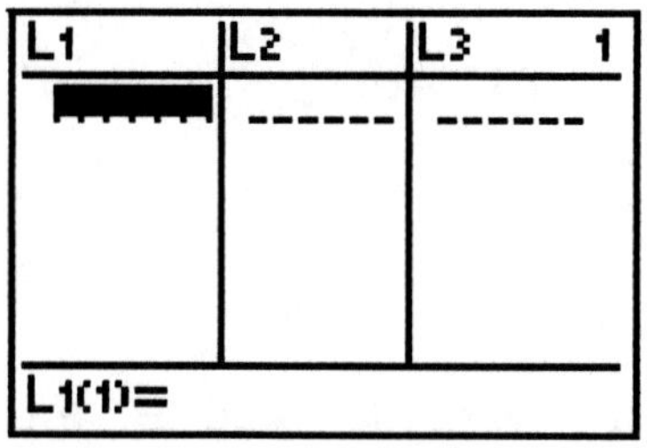

2. Enter the number of objects in **L1**.
3. Enter the distance from the floor to the bottom of the cup in **L2**.

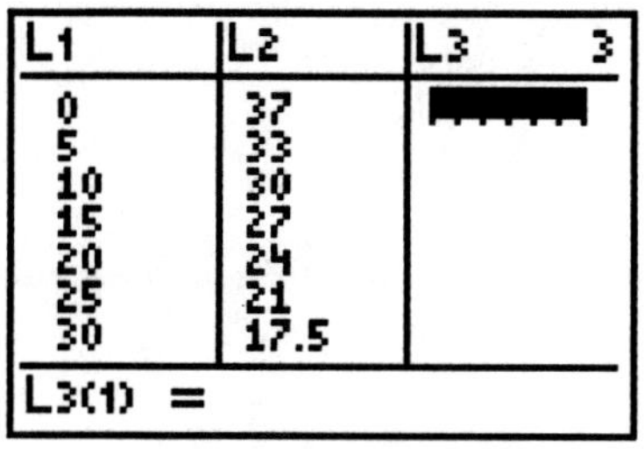

Setting up the window

1. Press [WINDOW] to set up the proper scale for the axes so that **ΔX** is **.5**.
2. Set the **Xmin** value by identifying the minimum value in **L1**. Choose a number that is less than the minimum.

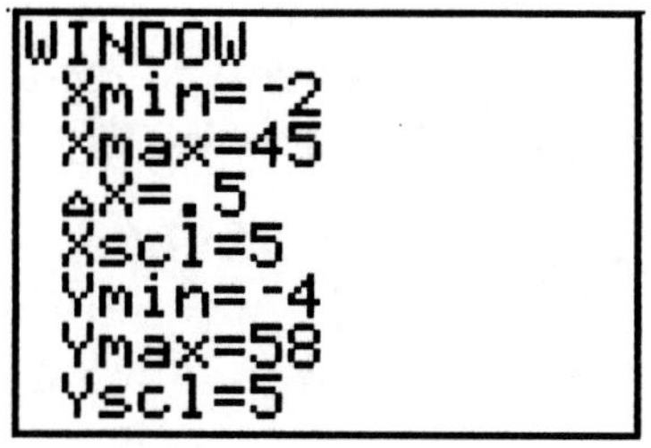

3. Set the **Xmax** value by identifying the maximum value in each list. Choose a number that is greater than the maximum. **Do Not Change the ΔX Value.** Set the **Xscl** to **5**.
4. Set the **Ymin** value by identifying the minimum value in **L2**. Choose a number that is less than the minimum.
5. Set the **Ymax** value by identifying the maximum value in **L2**. Choose a number that is greater than the maximum. Set the **Yscl** to **2**.

Graphing the data: Setting up a scatter plot

1. Press [2nd] [PLOT]. Select **1:Plot1** by pressing **1** or [ENTER].

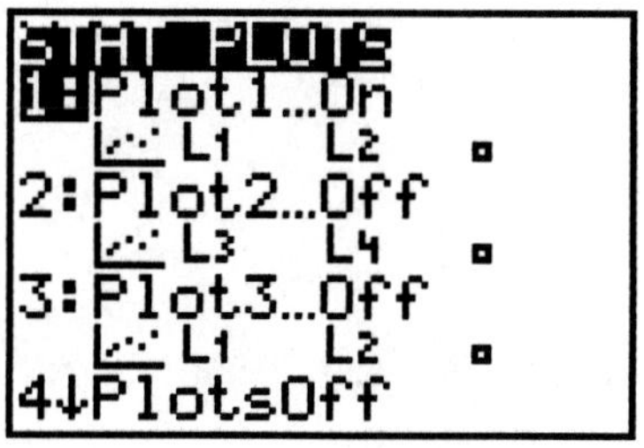

2. Set up the plot as shown by pressing [ENTER] [▼] [ENTER] [▼] [2nd] [STAT] **1:L1** [▼] [2nd] [STAT] **2:L2** [▼] [ENTER].

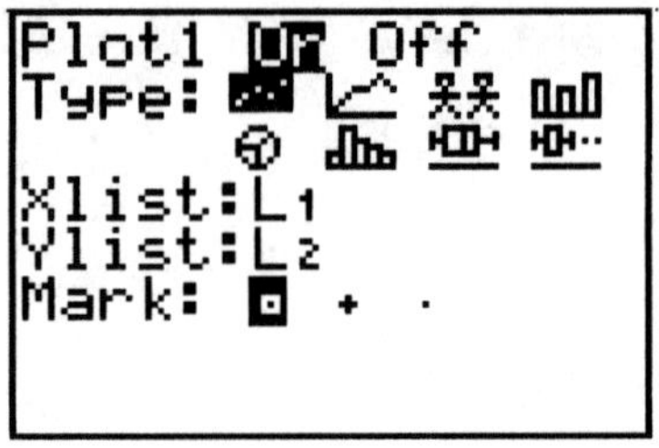

3. Press [GRAPH] to see the plot.

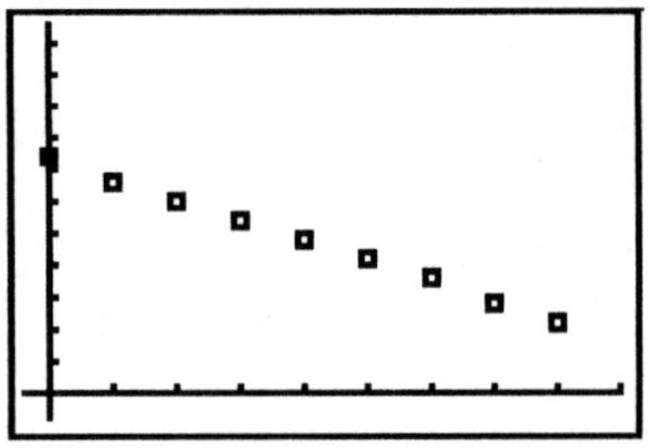

Analyzing the data

Finding a trend line

The data that you collected appears to be linear; therefore, you will find a linear equation for the line.

1. The *y*-intercept of a line is the point at which the line crosses the *y*-axis. The y-intercept of the trend line is the first value in **L2**. Record the *y*-intercept of the line on the **Data Collection and Analysis** page.

   Find a line of best fit using the **Manual-Fit** feature on the TI-73. **Manual-Fit** allows you to fit a line to plotted data on the Graph screen manually.

2. Press [2nd] [STAT] [◄] to move the cursor to the **CALC** menu.

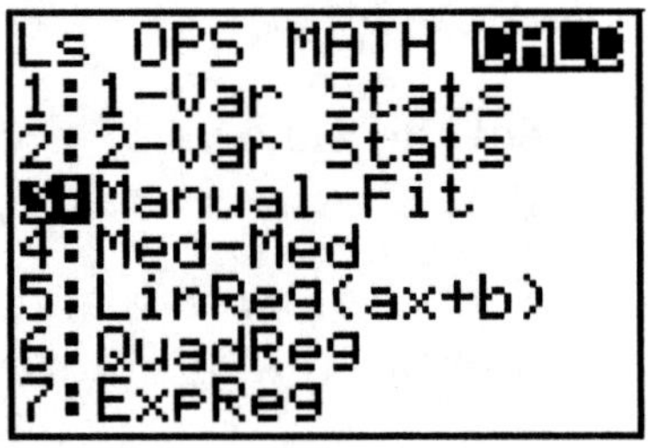

3. Select **3:Manual-Fit** by pressing **3**.

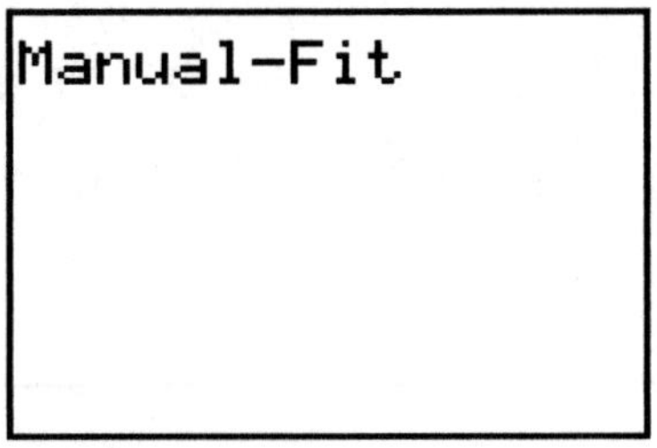

4. Press [2nd] [VARS]. Select **2:Y-Vars** by pressing **2**.

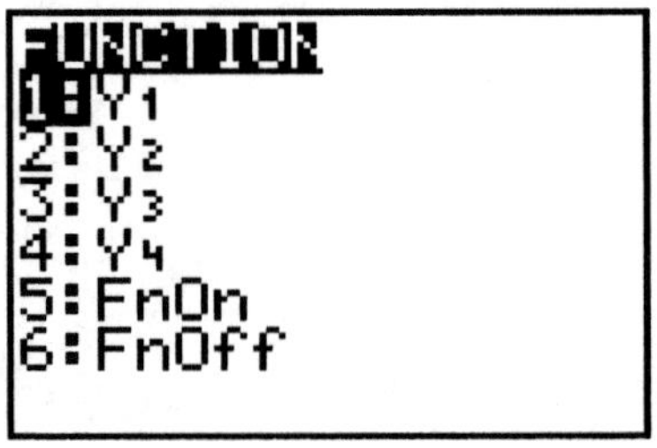

5. Select **1:Y1** by pressing **1** or [ENTER].

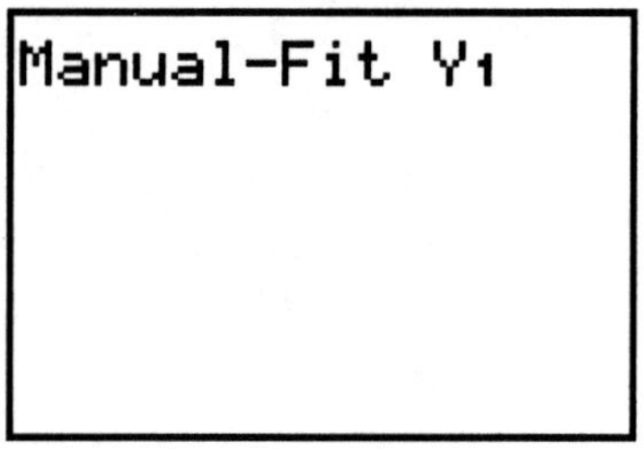

6. Press [ENTER] to perform the manual fit.

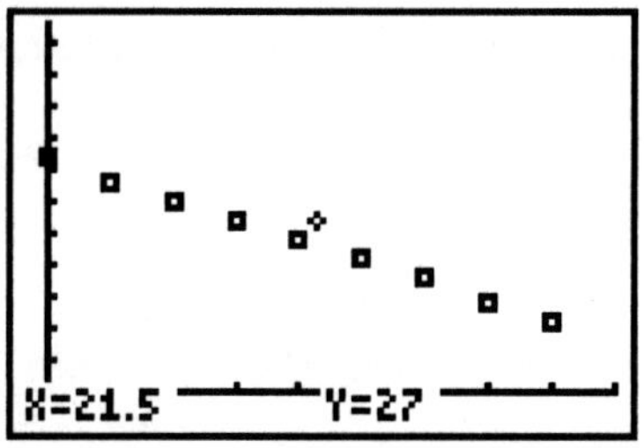

7. Use [▲] and [◄] to move the cursor to the *y*-intercept. (For this example, (0, 37).)

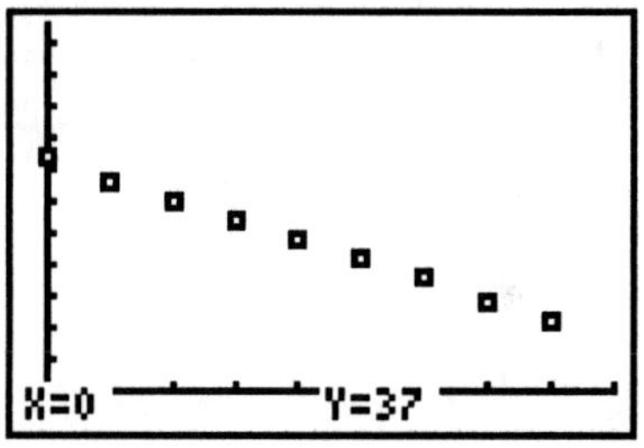

8. Press [ENTER] to make this point one point on the manual fit line.

9. Press [►] to extend a horizontal line across the screen.

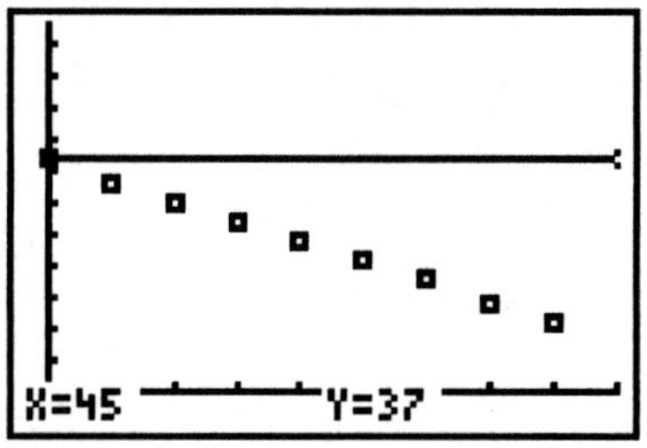

10. Press [▼] to adjust the slope of the line to match the data points.

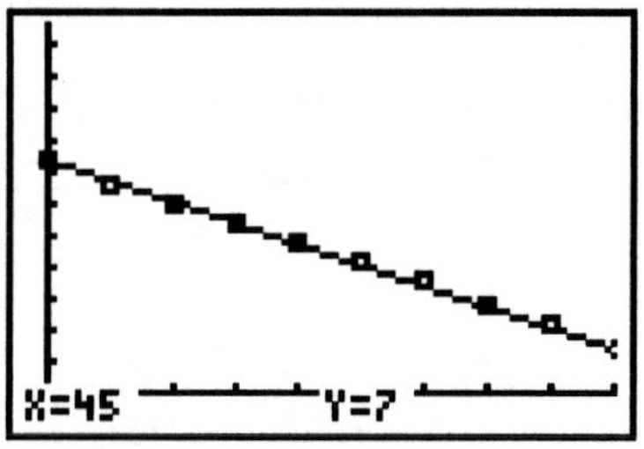

11. Press [ENTER] to anchor a second point on the manual fit line. (For this example, (45, 7).)

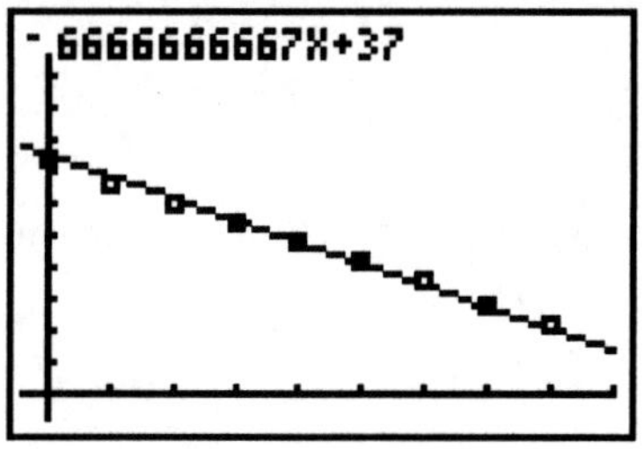

12. Use [▼] and [▲] to make adjustments to the slope. Use [◄] and [►] to make adjustments to the y-intercept. When you have found the line you feel best represents the data, press [ENTER] to save the manual fit line. The equation is pasted in **Y1**.

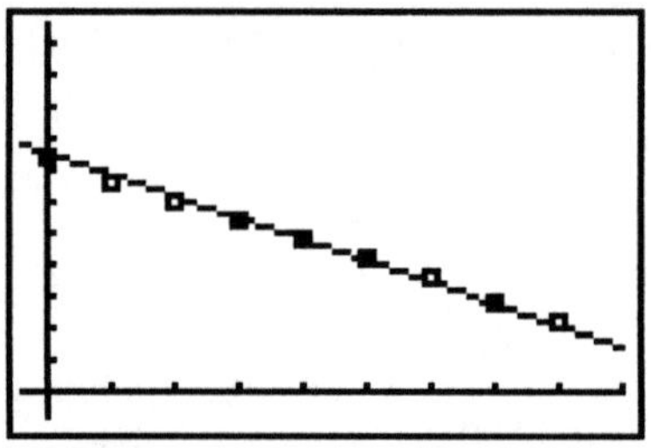

13. Press [Y=] to see the equation.

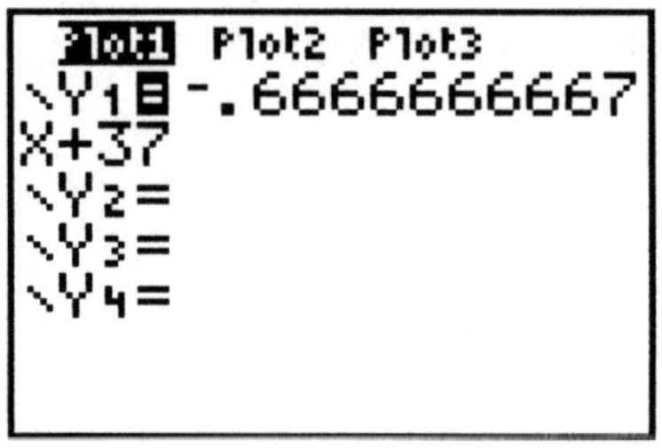

14. Record the slope and the equation of the line on the **Data Collection and Analysis** page.

Answer Part I questions 1-4 on the **Data Collection and Analysis** page.

Predicting the distance

You can predict the distance the cup will be from the floor based upon the number of objects you place in the cup. Use your model to determine the distance the cup is from the floor when 13 objects are added to the cup.

1. Press [TRACE].

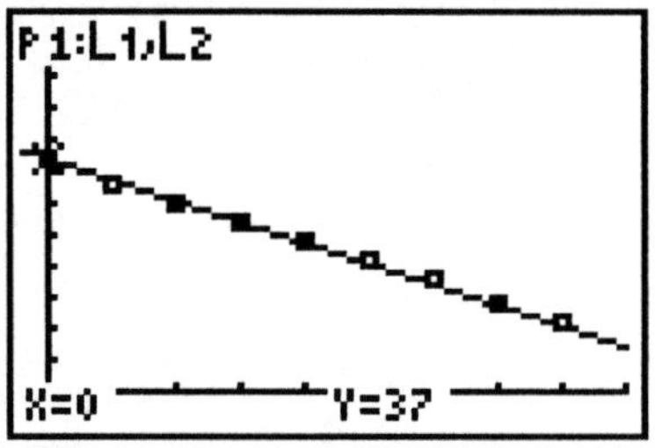

2. Press [▼] to get to **Y1**. Type **13** (the number of objects.)

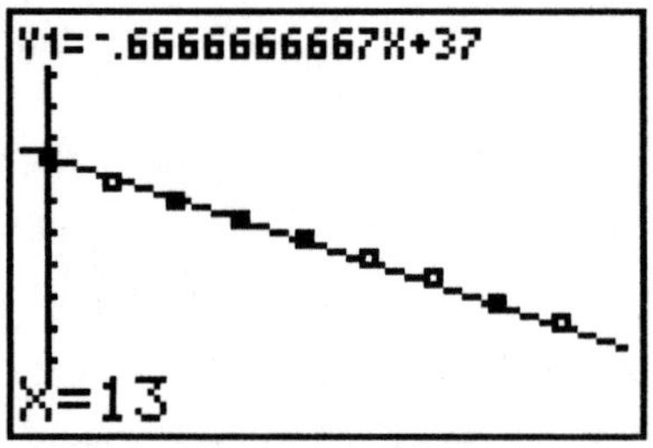

3. Press [ENTER]. The *x*-value represents the number of objects, and the *y*-value represents the distance the cup is from the floor.

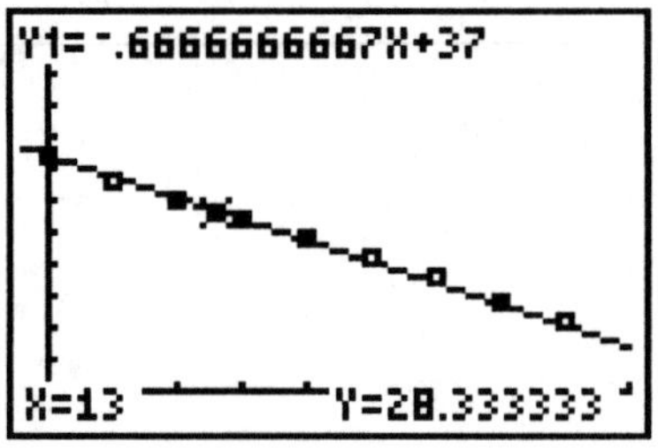

Answer Part I questions 5-6 on the **Data Collection and Analysis** page.

Predicting the number of objects

You can predict the number of objects in the cup based upon the distance the cup is from the floor. Use your model to determine the number of objects in the cup when the cup is 27 centimeters from the floor.

1. Press [Y=] and [▼] until you are in the first position for **Y2**. Type **27**.

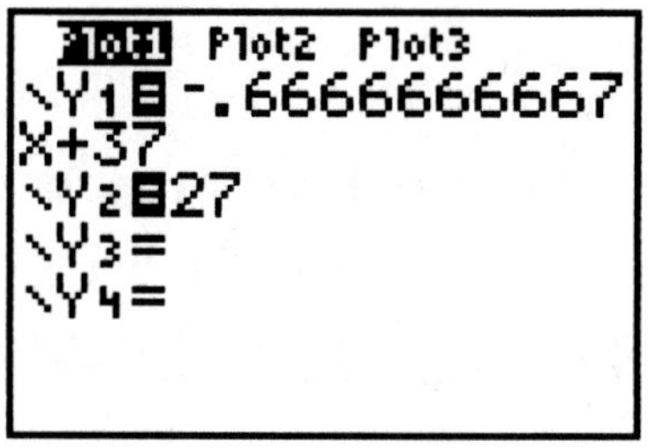

2. Press [GRAPH] to see the graph of the two intersecting lines.

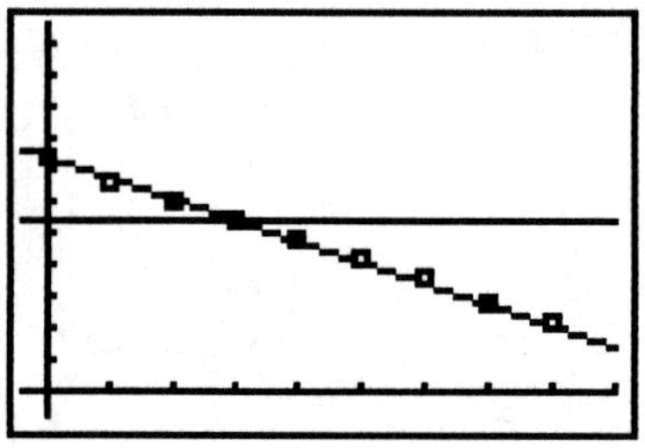

3. Draw a vertical line at the point of intersection. Press [DRAW]. Select **4:Vertical** by pressing **4**.

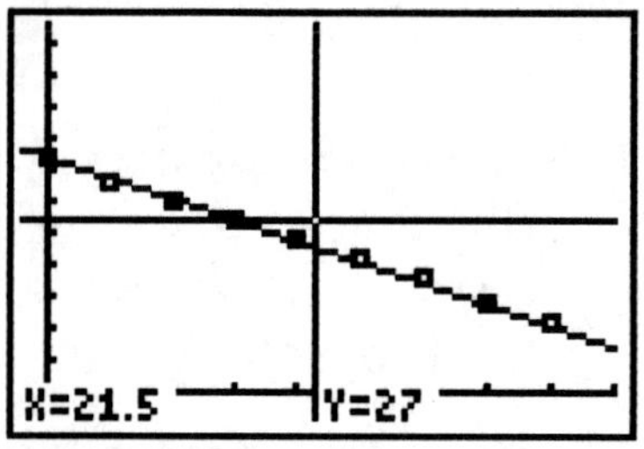

4. Use [◄] and [►] to move the vertical line until you reach the point of intersection.

   ***Note**: The x value is an estimate of the number of objects in the cup.*

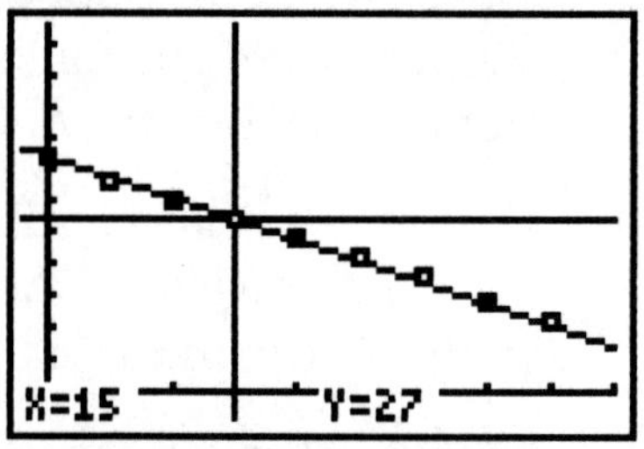

5. Use the Table to find the actual point of intersection. Press [2nd] [TBLSET]. Press [▼] [▼] [►] [ENTER] to set the Independent variable to **Ask**.

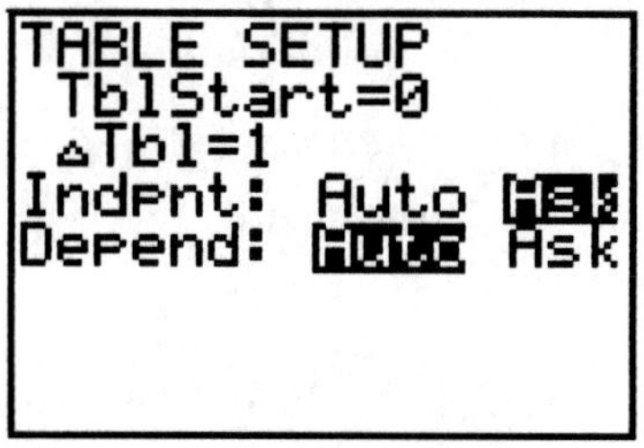

6. Press [2nd] [TABLE]. Enter *x* values and press [ENTER] until your *y* value is close to or equal to 27.

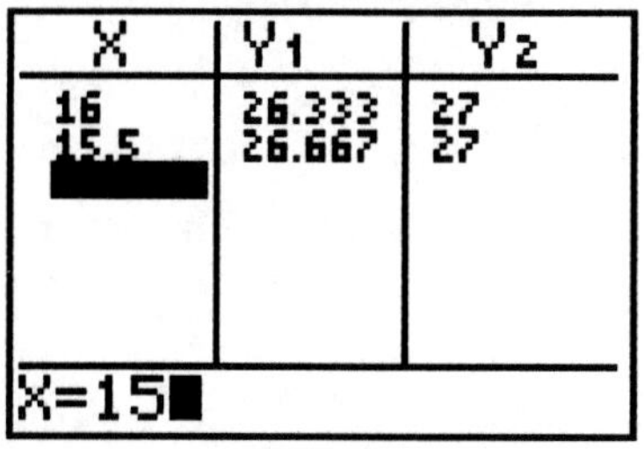

7. Press [▶] to examine the actual *y* value.

   ***Note**: Once you have entered seven x values, entering additional values for x will overwrite the seventh value.*

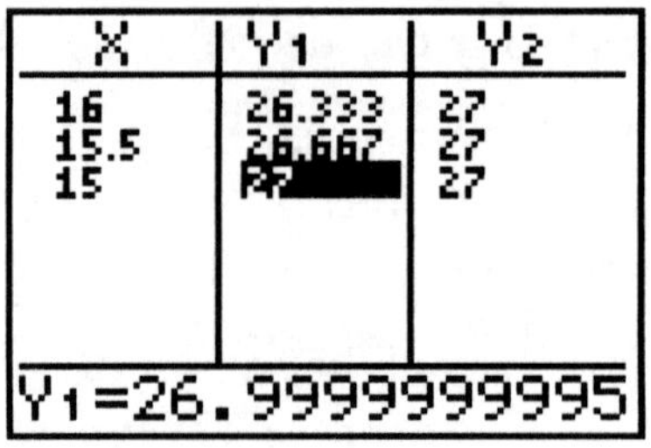

Answer Part I questions 7 - 8 on the **Data Collection and Analysis** page.

Collecting the Data — Part II

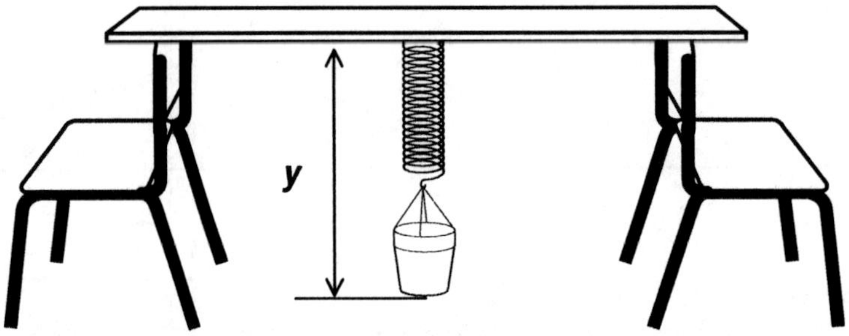

1. Remove all objects from the cup. Measure the length of the Slinky® and the cup. Place five of the objects that you are using in the cup. When the Slinky is stable, measure the length of the Slinky and the cup. Record the length in the table on the **Data Collection and Analysis** page.
2. Place five additional objects in the cup. When the Slinky is stable, measure the length of the Slinky and the cup. Record the length in the table on the **Data Collection and Analysis** page.
3. Continue placing additional objects in the cup in increments of five and measure the length of the Slinky and the cup. Record the lengths in the table on the **Data Collection and Analysis** page.
4. Measure the length of the cup with the paper clip. Record the length on the **Data Collection and Analysis** page.
5. Press [LIST]. Enter the length of the Slinky and the cup in **L3**.

   Use the following steps to calculate the length of the Slinky.

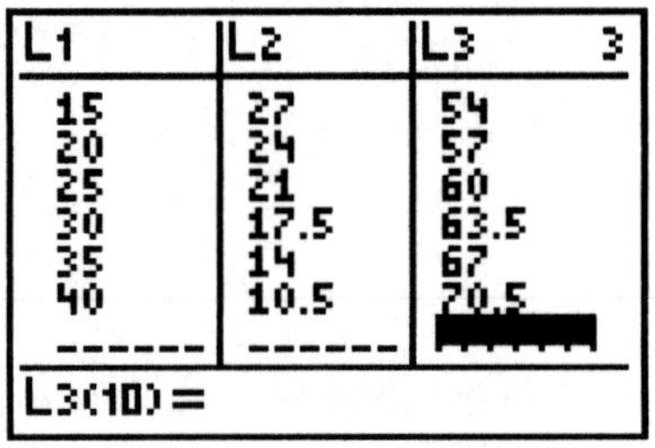

6. Press [▶] and [▲] to move the cursor to the top of **L4**.

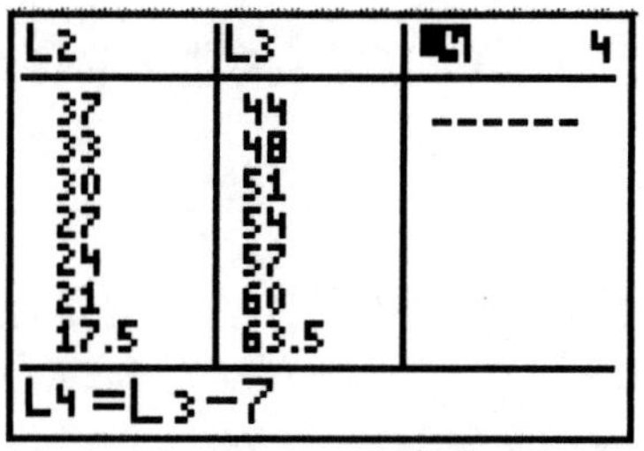

L2 L3 L4 4
37 33 30 27 24 21 17.5
44 48 51 54 57 60 63.5

L4 =L3-7

7. Press [2nd] [STAT] **3:L3** [−] the length of the cup and paper clip (recorded in Step 4.)

   ***Note**: For this example, the length of the cup and paper clip is 7 centimeters.*

8. Press [ENTER].

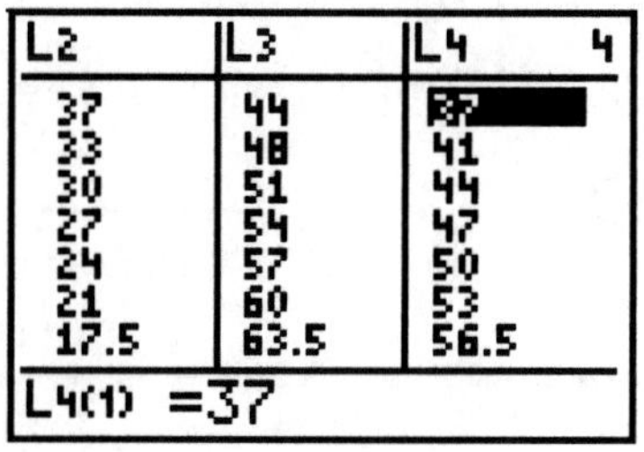

L2 L3 L4 4
37 33 30 27 24 21 17.5
44 48 51 54 57 60 63.5
41 44 47 50 53 56.5
L4(1) =37

9. Repeat the following Part I sections: **Setting up the window, Setting up a scatter plot**, and **Finding a trend line**, using the data for the number of objects (**L1**) and the length of the Slinky® and cup (**L3**). When performing the manual fit, use **Y2** instead of **Y1**. Be sure to turn off **Y1** by pressing [Y=] [◀] [ENTER] before viewing the **StatPlot**.

10. Repeat step 9 using the data for the number of objects (**L1**) and length of the Slinky (**L4**). When performing the manual fit, use **Y3** instead of **Y1**. Be sure to turn off **Y2** by pressing [Y=] [◀] [▼] [ENTER] before viewing the **StatPlot**.

Use equations **Y1**, **Y2**, and **Y3** to answer Part II questions 1 through 6 on the **Data Collection and Analysis** page.

Data Collection and Analysis

Name ______________________

Date ______________________

Activity 10: That's a Stretch

Collecting the data — Part I

Record your data from Part I in the table below.

Number of objects in cup	Distance from floor to bottom of cup (cm)
0	
5	
10	
15	
20	
25	
30	
35	
40	

Analyzing the data — Part I

The *y*-intercept is: ____________.

Slope = ________ Equation of Line **Y1**: ____________________

Use your equation of line (**Y1**) to answer questions 1 through 8.

1. What is the *independent variable* for this activity? ______________________

2. What is the *dependent variable* for this activity? ______________________

3. Explain what the *y*-intercept represents.

____

____

4. Explain what the *slope* represents.

____

____

5. Use your equation to find the distance from the floor to the bottom of the cup if 13 objects were placed in the cup. ____________________

6. Actually add 13 objects to the cup and measure the distance the cup is above the floor. How does this value compare to the value predicted in question 5?

____

____

____

____

7. Using the data that you collected, determine how many objects were used if the distance measured 27 centimeters. ____________________

8. Jennifer did this activity with 40 pennies and Mustafa did this activity with 40 small candies (M&M's® or Skittles®). Draw a sketch of the lines produced by Jennifer and Mustafa on the same set of axes. Label the axes. Identify which line represents Jennifer's data and which line represents Mustafa's data.

a. Which person had a line with the smaller *slope*?

____

b. Which person had a line with the greater *y*-intercept?

____

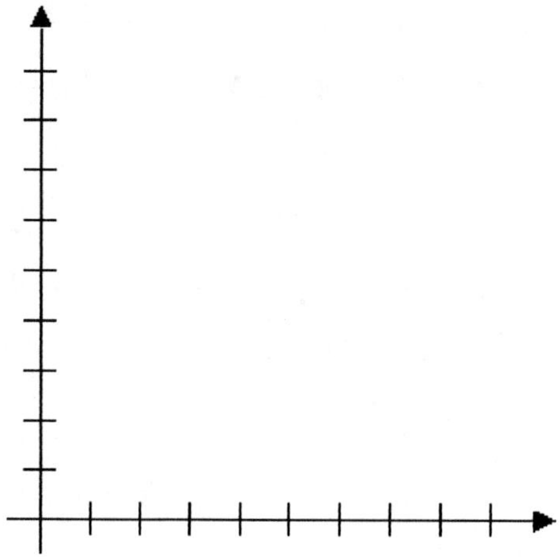

Collecting the data — Part II

Record your data from Part II in the table below.

Number of objects in cup	Length of Slinky® and cup (cm)	Length of Slinky (cm)
0		
5		
10		
15		
20		
25		
30		
35		
40		

Analyzing the data — Part II

Length of Cup and Paper Clip = ________________

For Slinky and cup:

The *y*-intercept is: ___________.

Slope = ______ Equation of Line **Y2**: ____________________

For length of Slinky:

The *y*-intercept is: ___________.

Slope = ________ Equation of Line **Y3**: ____________________

1. How do the *slopes* of the lines in equations **Y1**, **Y2**, and **Y3** compare?

____

____

2. What is the meaning of the *slope* in equations **Y2** and **Y3**?

Equation **Y2**: ____

____

Equation **Y3**: ____

____

3. Explain the meaning of the *y*-intercept in equations **Y2** and **Y3**.

Equation **Y2**: ____

____

Equation **Y3**: ____

____

4. How far would the Slinky® stretch if 13 objects were placed in the cup?

____

Note: *See the* ***Predicting the distance*** *section of Part I for instructions on how to do this.*

5. How many objects would it take to stretch the Slinky a distance of 75 centimeters?

____

Note: *See the* ***Predicting the number of objects*** *section of Part I for instructions on how to do this.*

6. Repeat question number 6 in Part I, but this time sketch a graph for the length of the Slinky. Record you answers to **a** and **b** below.

a. ____________________

b. ____________________

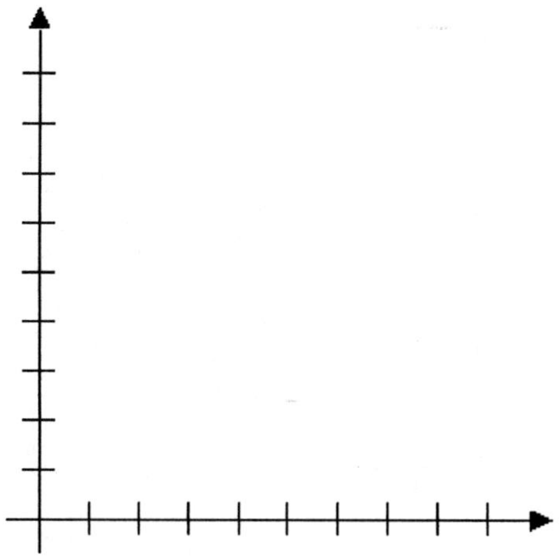

Teacher Notes

Activity 10

That's a Stretch

Objectives

- To determine the relationship between the stretch of a spring and the number of weights in a cup suspended from the spring
- To find the *y* value of a function, given the *x* value
- To find the x value of a function, given the y value
- To use technology to find a best fit line
- To use technology to plot a set of ordered pairs

Materials

- TI-73 graphing device
- Slinky® cut in half, one per group
- Small bathroom paper cups or film containers, one per group
- Marbles, pennies, or other small objects such as cubes, at least 40 per group
- Large paper clips, one per group
- Meter stick, one per group

Preparation

- You can suspend the meter stick across two desks or two chairs.
- You can use marbles, pennies, small cubes, or candy for objects to place in the cup.
- This activity explores both positive and negative slopes. Part II of the activity allows students to examine the *y*-intercept of a line.

Answers to Data Collection and Analysis

Collecting the data

- Sample data, Part I:

Number of objects in cup	Distance from floor to bottom of cup (cm)
0	37
5	33
10	30
15	27
20	24
25	21
30	17.5
35	14
40	10.5

- Sample data, Part II:

Number of objects in cup	Length of Slinky® and cup (cm)	Length of Slinky (cm)
0	44	37
5	48	41
10	51	44
15	54	47
20	57	50
25	60	53
30	63.5	56.5
35	67	60
40	70.5	63.5

Analyzing the data — Part I

Use your equation of line (**Y1**) to answer questions 1 through 7.

1. What is the *independent variable* for this activity?

The independent variable for this activity is number of objects.

2. What is the *dependent variable* for this activity?

The dependent variable for this activity is the distance from the floor to the bottom of the cup.

3. Explain what the *y*-intercept represents.

The y-intercept is the distance, in centimeters, from the bottom of the cup to the floor with zero objects in the cup.

4. Explain what the *slope* represents.

The slope represents the number of centimeters that the distance from the bottom of the cup to the floor changes each time an object is added to the cup.

5. Use your equation to find the distance from the floor to the bottom of the cup if 13 objects were placed in the cup.

For the sample data, the distance is 28.33 cm.

6. Actually add 13 objects to the cup and measure the distance the cup is above the floor. How does this value compare to the value predicted in question 5?

Answers will vary. The values should be close.

7. Using the data you collected, determine how many objects were used if the distance measured 27 centimeters?

The TI-73 returns a value of 15.345; however, the answer must be an integer. Therefore, the value is at least 16 objects. Check and discuss students' answers.

8. Jennifer did the activity with 40 pennies and Mustafa did the activity with 40 small candies (M&M's® or Skittles®). Draw a sketch of the lines produced by Jennifer and Mustafa on the same set of axes. Label the axes. Identify which line represents Jennifer's data and which line represents Mustafa's data.

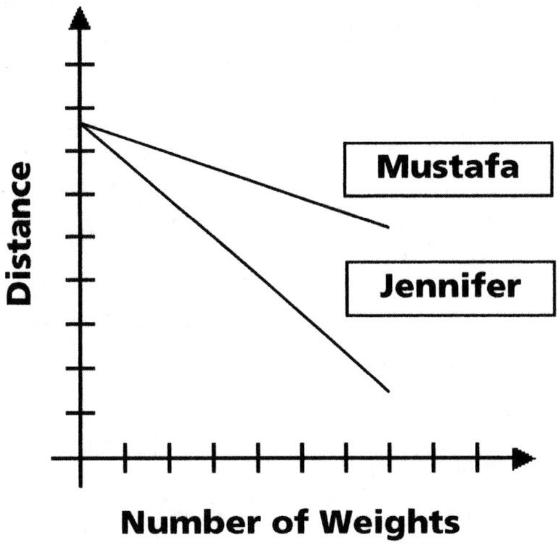

a. Which person had a line with the smaller slope?

Mustafa

b. Which person had a line with the greater *y*-intercept?

They have the same y-intercept.

Analyzing the data — Part II

1. How do the *slopes* of the lines in equations **Y1**, **Y2**, and **Y3** compare?

   The slopes of all of the lines are equal in absolute value. However, slopes of equations Y2 and Y3 are positive while the slope of equation Y1 is negative.

2. What is the meaning of the *slope* in equations **Y2** and **Y3**?

   Equation **Y2**: *The number of centimeters that the length of the spring including the cup increases each time an object is added to the cup.*

   Equation **Y3**: *The number of centimeters that the length of the spring increases each time an object is added to the cup.*

3. Explain the meaning of the *y*-intercept in equations **Y2** and **Y3**.

   Equation **Y2**: *The initial length, in centimeters, of the spring including the cup.*

   Equation **Y3**: *The initial length, in centimeters, of the spring.*

4. How far would the Slinky® stretch if 13 objects were placed in the cup?

   Based upon the sample data, 52.7 cm.

5. How many objects would it take to stretch the Slinky a distance of 75 centimeters?

   Based upon the sample data, 46.6 objects. Since only whole objects can be added, a reasonable value would be 47.

6. Repeat question number 6 in Part I, but this time sketch a graph for the length of Slinky®. Record your answers to **a** and **b** below.

   a. *Mustafa had the graph with the smaller slope.*

   b. *They have the same y-intercept.*

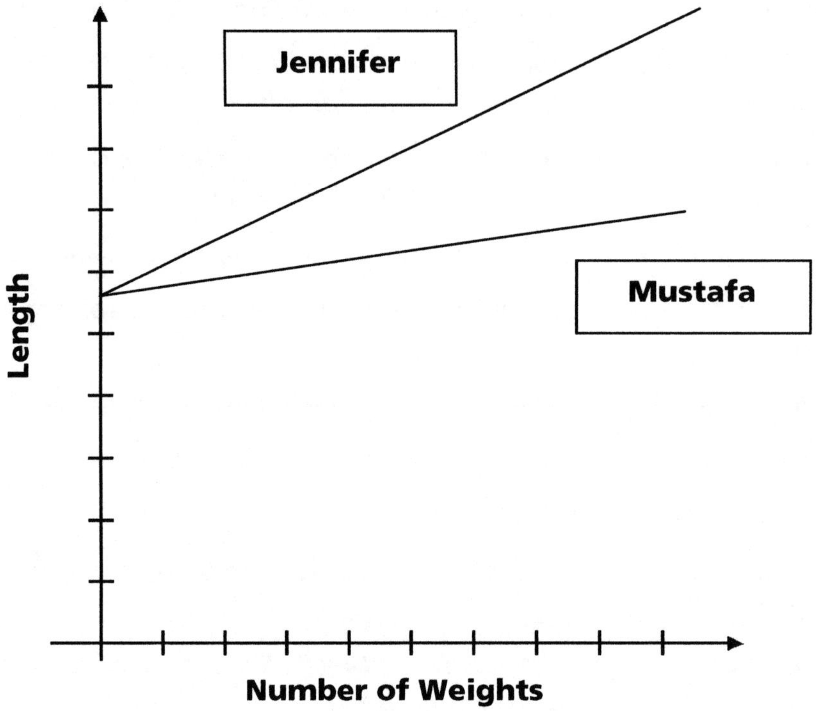

EXPLORATIONS

Activity 11

Get Your Numbers in Shape

Objectives

- To use technology to explore patterns
- To use inductive reasoning to make conjectures about patterns
- To use technology to produce a sequence
- To find the *x* value of a function, given the *y* value
- To find the *y* value of a function, given the *x* value
- To find a linear or a quadratic equation for a given pattern

Materials

- TI-73 graphing device
- Small cubes or candy

Introduction

Study the pictures below and find the next shape in each of the patterns shown.

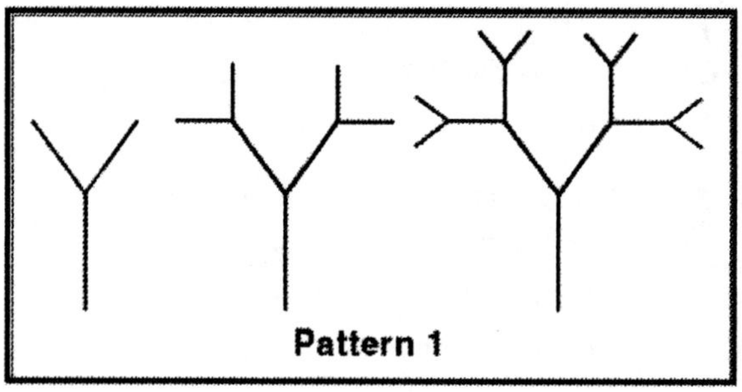

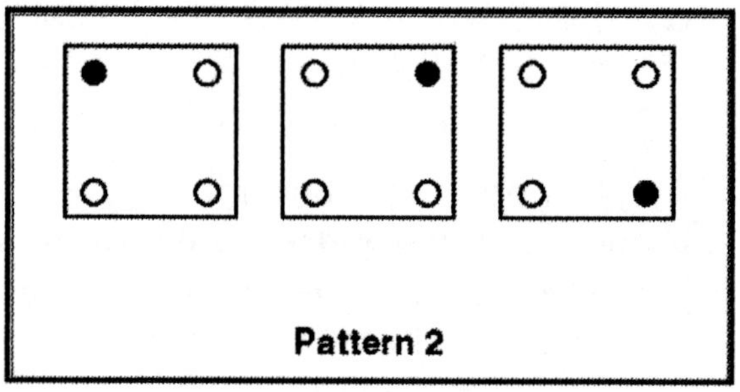

The process that you have used to find the next shape in each pattern is called *inductive reasoning*. You have used this type of reasoning since you were a baby. You learned how to eat using the proper utensils and how to turn on the television by observing others and drawing conclusions. After several trials, you perfected your skills. Inductive reasoning allows you to make generalizations based on a pattern of specific examples or past events. These generalizations are called *conjectures* or *hypotheses*. Mathematicians and scientists use inductive reasoning to make discoveries and develop formulas based on their discoveries.

Mathematicians have used number patterns to describe a variety of phenomena in nature. For example, a famous pattern that is found throughout nature is the *Fibonacci sequence* (see Activities 4 and 8.) In this activity, you will investigate a variety of sequences.

Problem

How do you use inductive reasoning to find a formula for the patterns that describe an array of objects or a sequence of numbers?

Collecting the data

A *triangular number* is a number that can be represented by a triangular arrangement of objects.

A *square number* is a number that can be represented by a square arrangement of objects. Numbers that correspond to geometric figures are called *figurate numbers*. The diagrams below illustrate triangular numbers and square numbers.

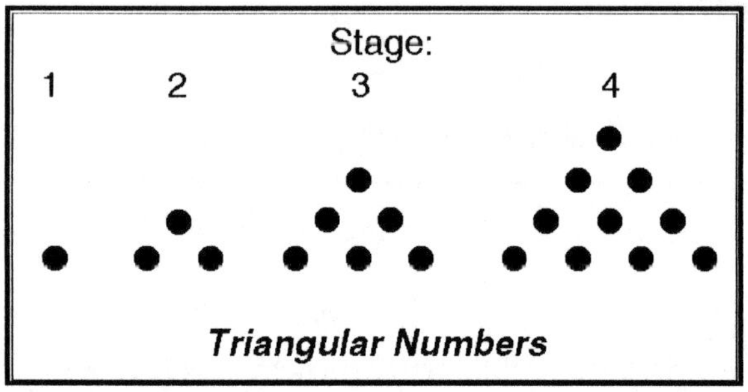

Triangular Numbers

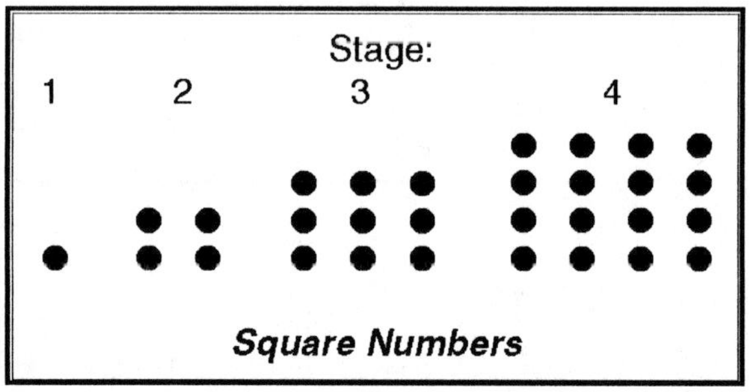

Square Numbers

Use the objects that your teacher will give you to find the next three stages for the triangular numbers. Complete the table on the **Data Collection and Analysis** page.

Do you notice a pattern in the table? Suppose you needed to know the 100th triangular number. You could continue to form the pattern using your objects, however this could become tedious and time-consuming. If you knew a rule or *function* for calculating any term in this sequence, you could find the 100th term of the sequence with relative ease.

You will use the *finite differences* technique to find a function for this sequence. This method requires that you find the difference between successive values in the sequence. If the first differences are equal, then the function or rule that describes the sequence is *linear.* If the second differences are equal, then the function or rule that describes the sequence is *quadratic*. You can use the TI-73 to find the differences.

Setting up the TI-73

Before starting your data collection, make sure that the TI-73 has the STAT PLOTS turned OFF, Y= functions turned OFF or cleared, the MODE and FORMAT set to their defaults, and the lists cleared. See the Appendix for a detailed description of the general setup steps.

Entering the data in the TI-73

Enter the stage number data in **L1** and the value in **L2**. You will use the *sequence* command on the TI-73 to enter the data in **L1**.

1. Press [LIST].

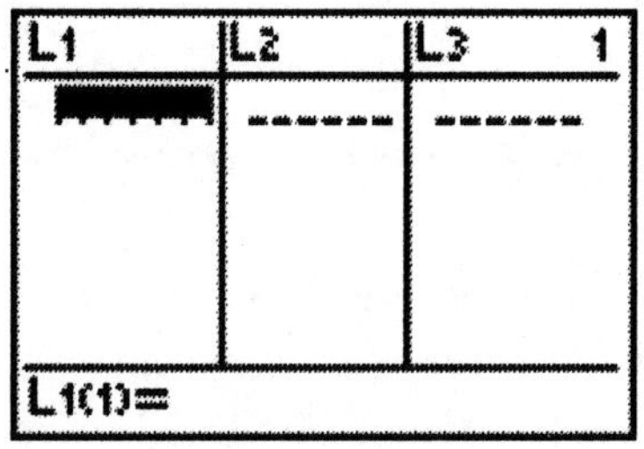

2. Press [▲] to highlight **L1**.

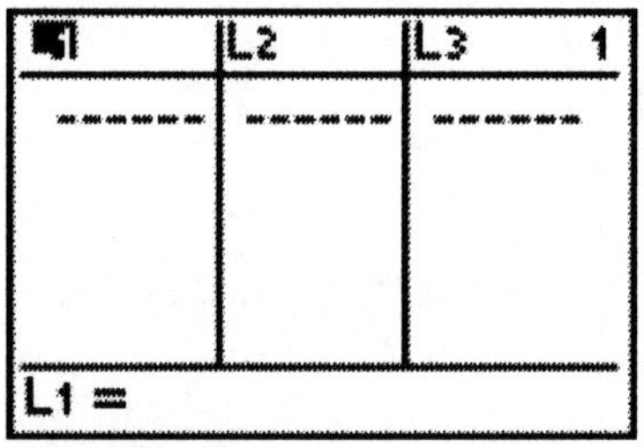

3. Press [2nd] [STAT] [▶] to move the cursor to the **OPS** menu.

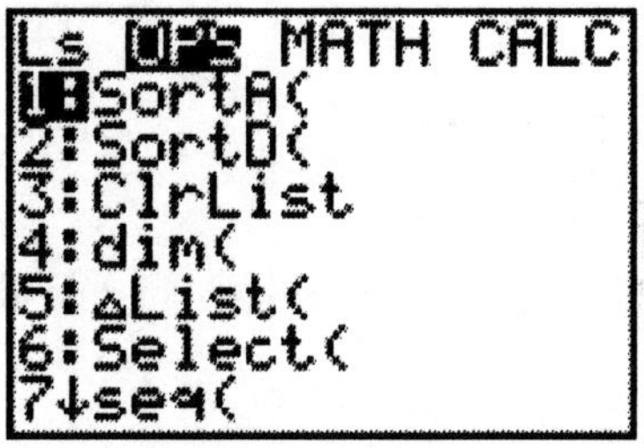

4. Select **7:seq(** by pressing **7**.

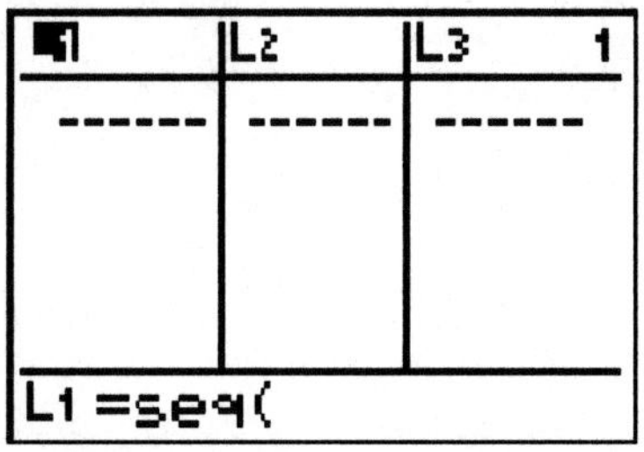

5. Press [x] [,] [x] [,] **1** [,] **8** [,] **1** [)] to complete the command.

   ***Note**: The components of the sequence command are **seq(** expression or formula, variable, beginning value, ending value, increment).*

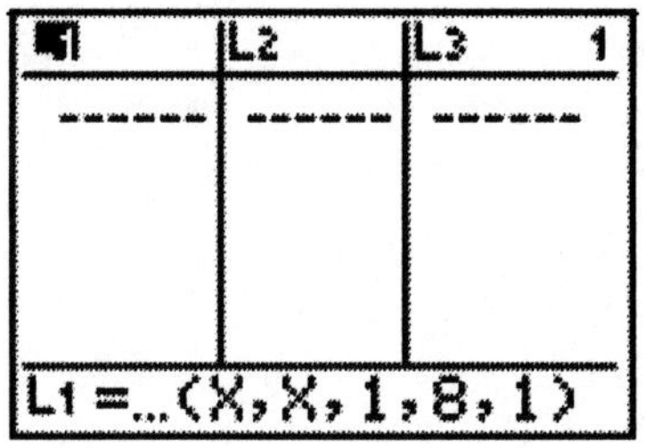

6. Press [ENTER]. The sequence of numbers from 1 to 8 will be placed in **L1**.

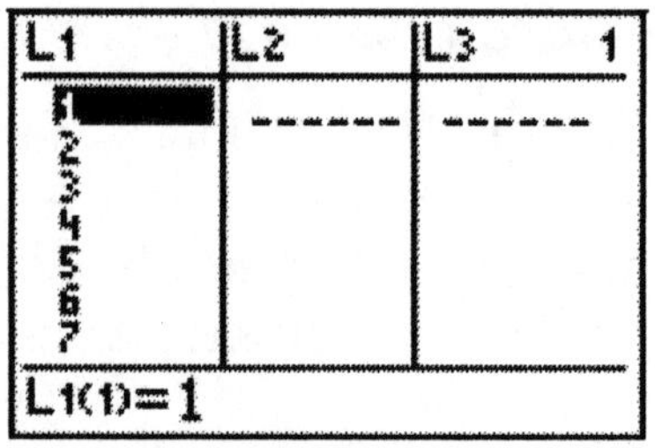

7. Enter the values for the first eight stages of the triangular numbers in **L2**.

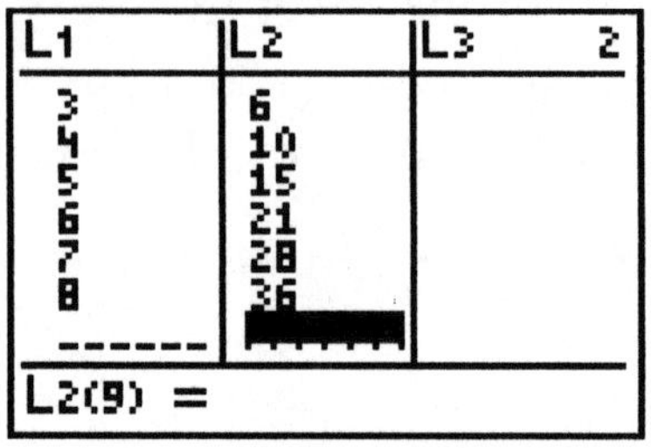

8. Find the first differences. Press [▶] [▲] to move the cursor to the top of **L3**.

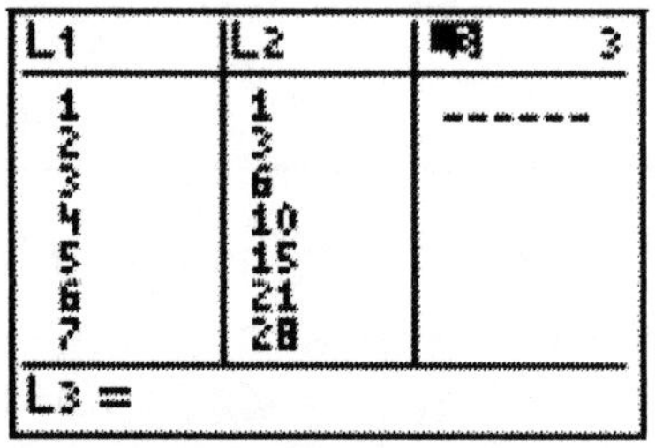

9. Press [2nd] [STAT] [▶] to move the cursor to the **OPS** menu.

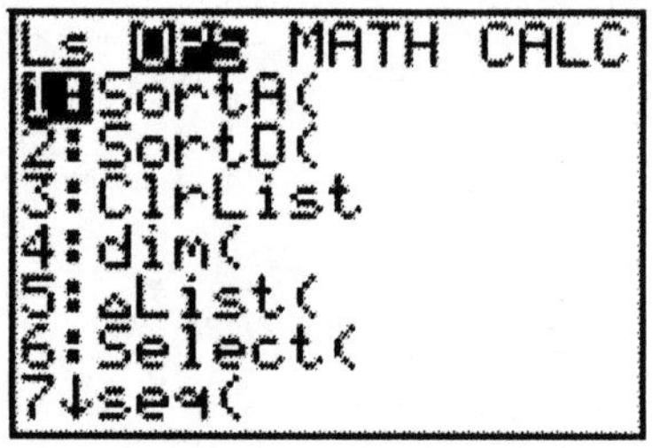

10. Select **5:ΔList(** by pressing **5**.

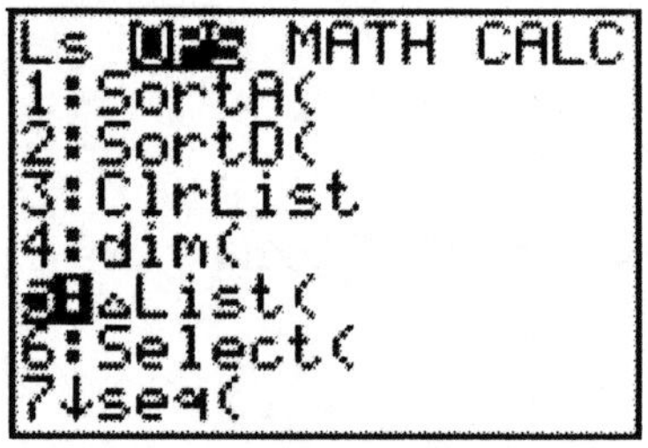

11. Press [2nd] [STAT] **2:L2** [)].

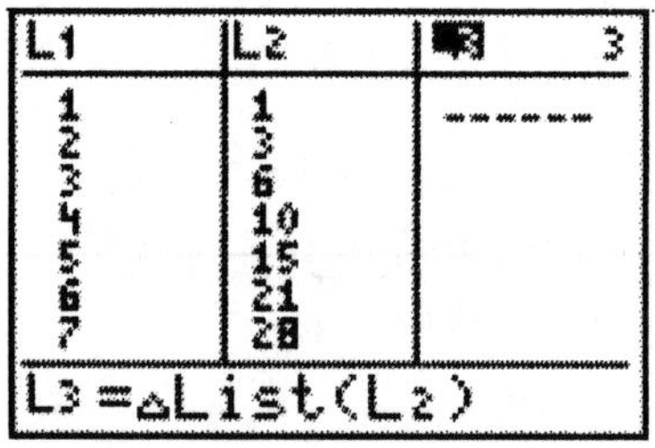

12. Press [ENTER] to see the difference between the successive terms in **L2**.

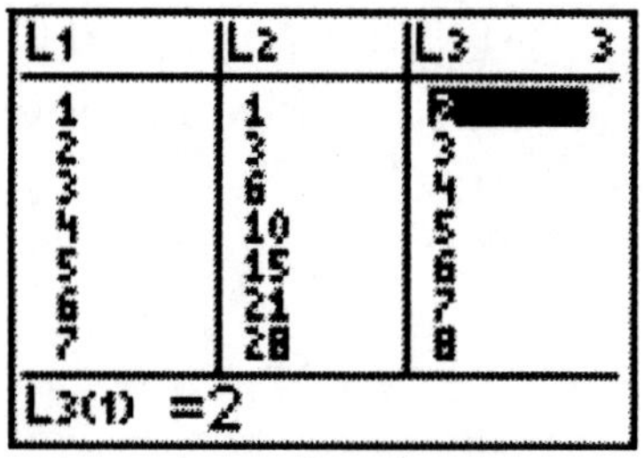

The numbers in **L3** represent the first difference of the sequence. Observe that the values are not equal. Therefore, the function that describes this sequence is not *linear*.

Term Number	1	2	3	4	5	6	7	8
Value	1	3	6	10	15	21	28	36

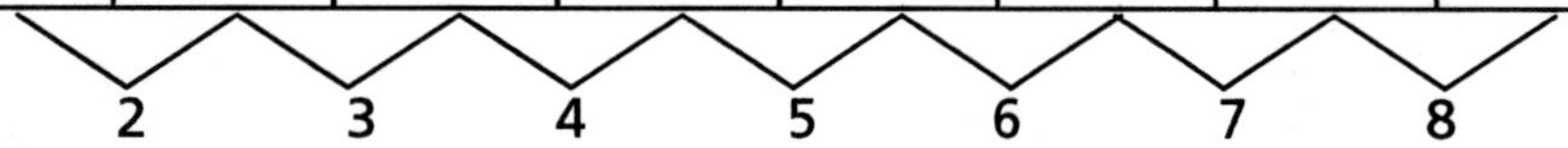

13. Find the second differences. Press [▶] [▲] to move the cursor to the top of **L4**.

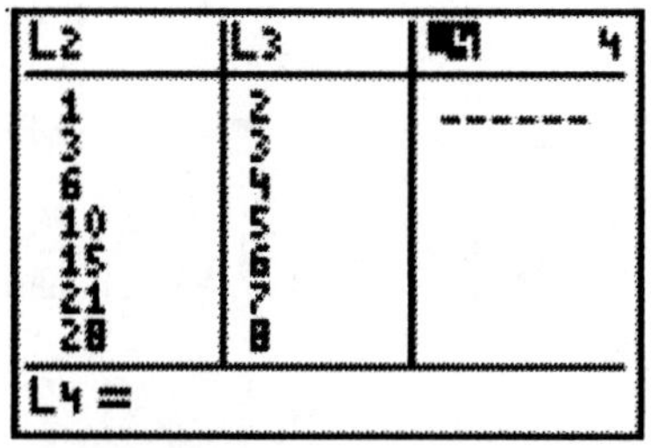

14. Press [2nd] [STAT] [▶] to move the cursor to the **OPS** menu.

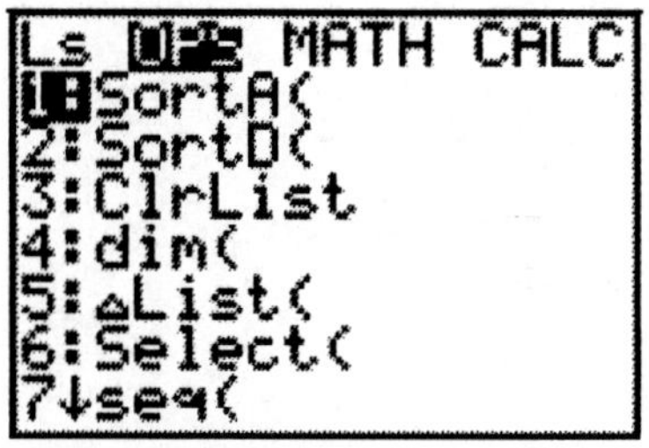

15. Select **5:ΔList(** by pressing **5**.

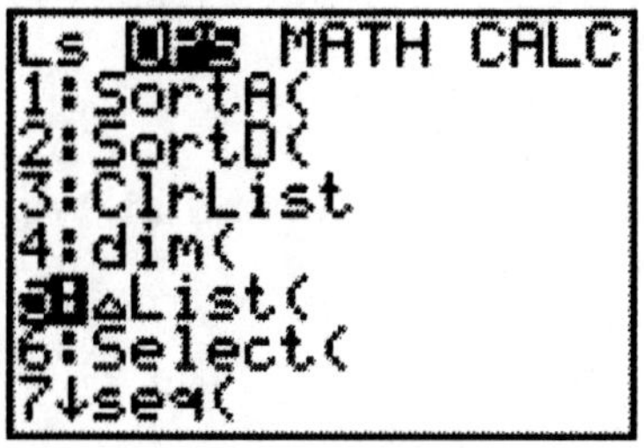

16. Press [2nd] [STAT] **3:L3** [)].

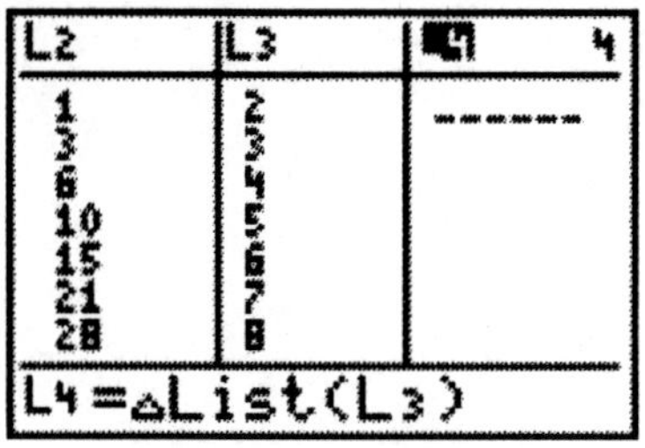

17. Press [ENTER] to see the difference between the successive terms in **L3**.

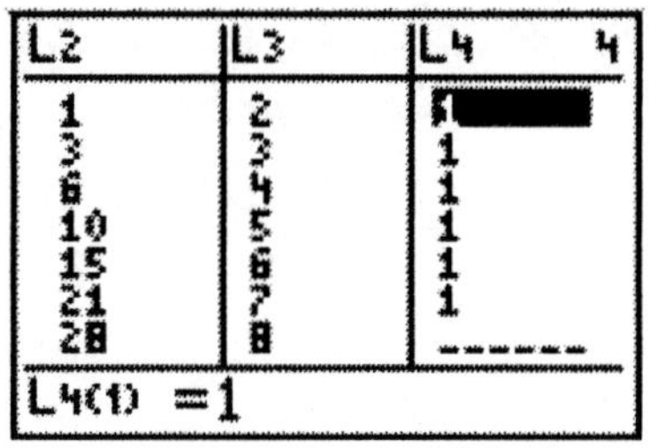

Since the second differences are equal, the function that describes this sequence is *quadratic.*

Graphing the data: Setting up a scatter plot

Plot the data using **L1** and **L2**.

1. Press [2nd] [PLOT]. Select **1:Plot1** by pressing **1** or [ENTER].

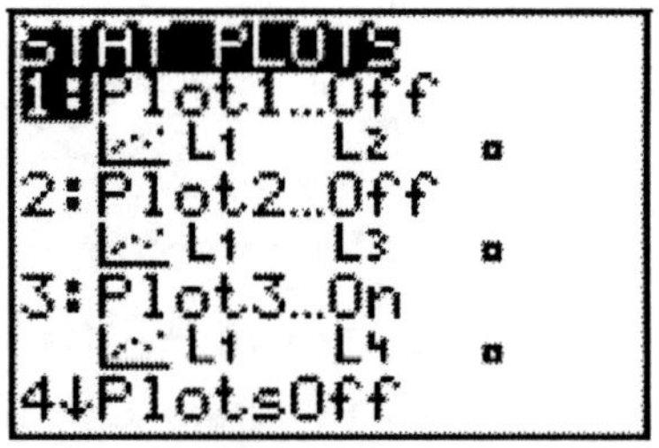

2. Set up the plot by pressing [ENTER] [▼] [ENTER] [▼] [2nd] [STAT] **1:L1** [▼] [2nd] [STAT] **2:L2** [▼] [ENTER].

3. Press [ZOOM].

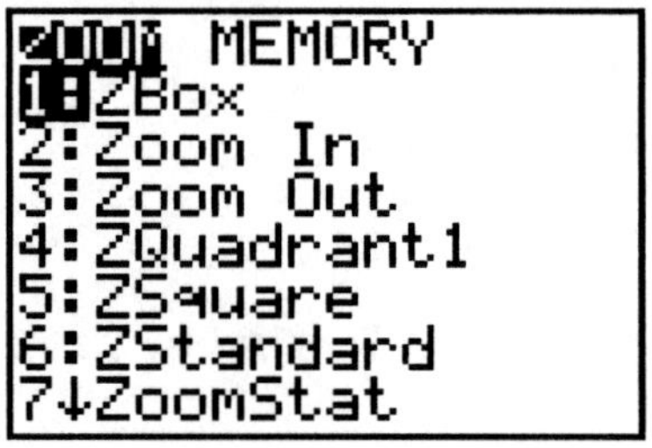

4. Select **7:ZoomStat** by pressing **7**.

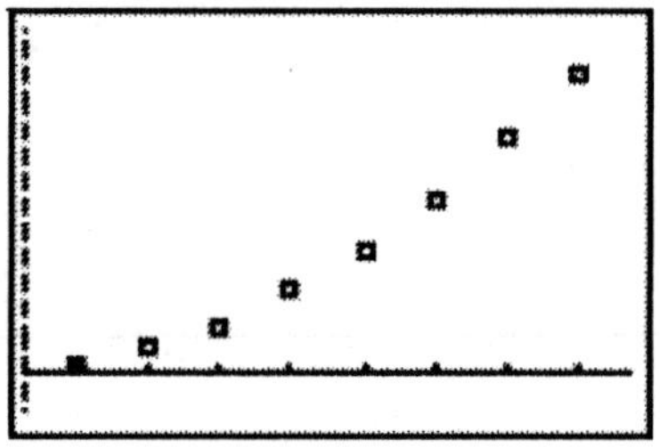

Analyzing the data: Finding a trend line

Use a quadratic regression to find the quadratic function that describes this sequence.

1. Press [2nd] [STAT] [◄] to move the cursor to the **CALC** menu.

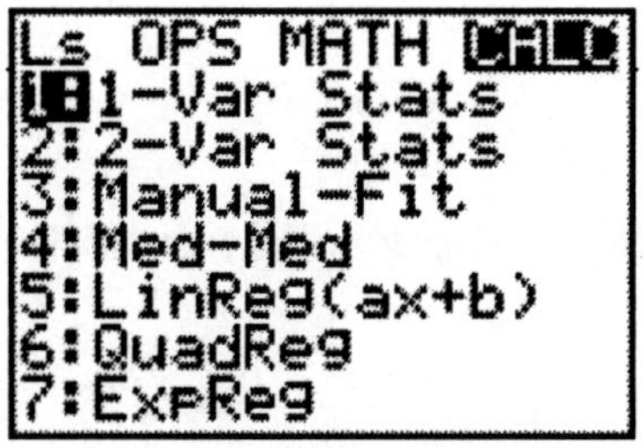

2. Select **6:QuadReg** by pressing **6**.

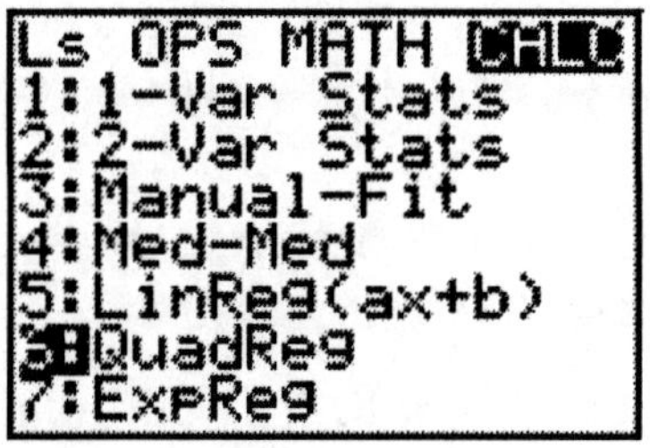

3. Press [2nd] [STAT] **1:L1** [,] [2nd] [STAT] **2:L2** [,].

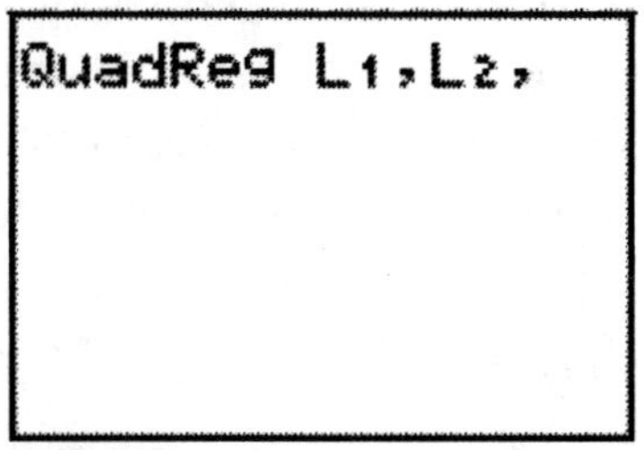

4. Press [2nd] [VARS]. Select **2:Y-VARS** by pressing **2**.

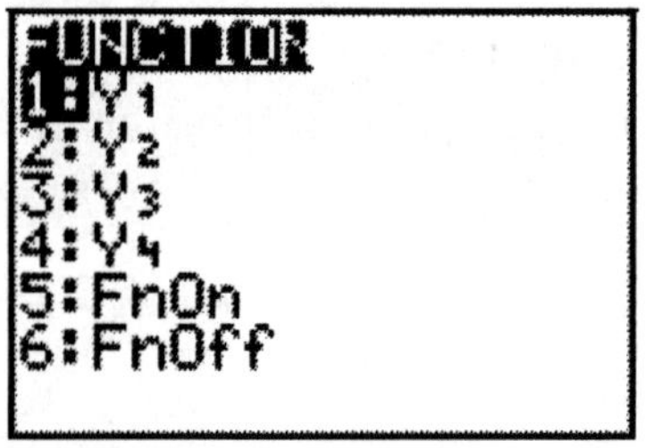

5. Select **1:Y1** by pressing **1** or [ENTER].

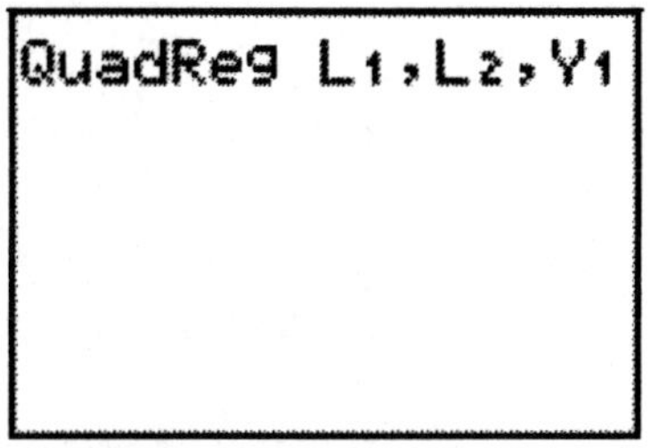

6. Press [ENTER] to see the function.

   The function that describes the sequence for triangular numbers has been pasted in **Y1**.

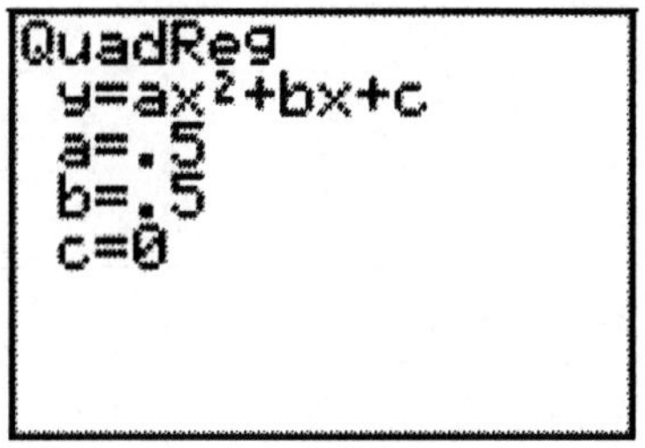

7. Press [Y=] to see the equation.

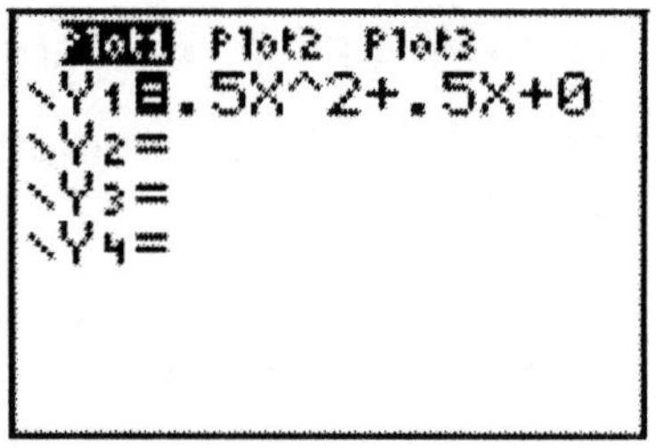

8. Press [GRAPH] to see the graph of the function.

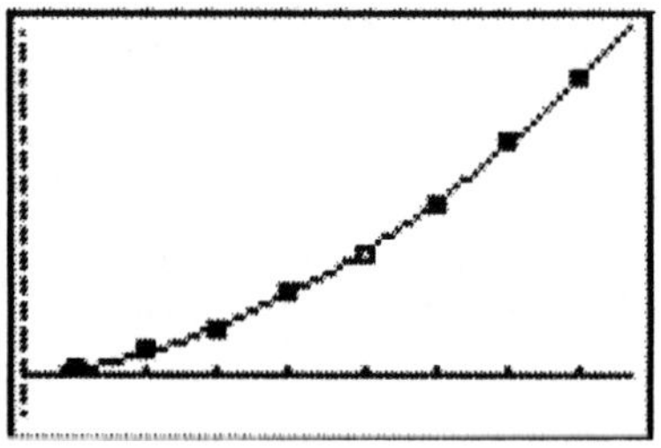

Use the function to answer Part I questions 1 through 4 on the **Data Collection and Analysis** page.

The formula for triangular numbers is a sequence. A *sequence* is an ordered list of numbers. You have graphed the formula in function mode. In function mode, values such as 1.5, 2.3, and 4.7 can be evaluated. For a sequence, however, the domain can only be integers. Therefore, the plot of the data is a better representation of the triangular numbers than is the graph of the function.

Finding the formula, a different approach

Another way to find a formula for the *triangular numbers* is illustrated below. Examine the diagram and table below for triangular numbers. The black dots represent a triangular number, and the white dots represent a second triangular number.

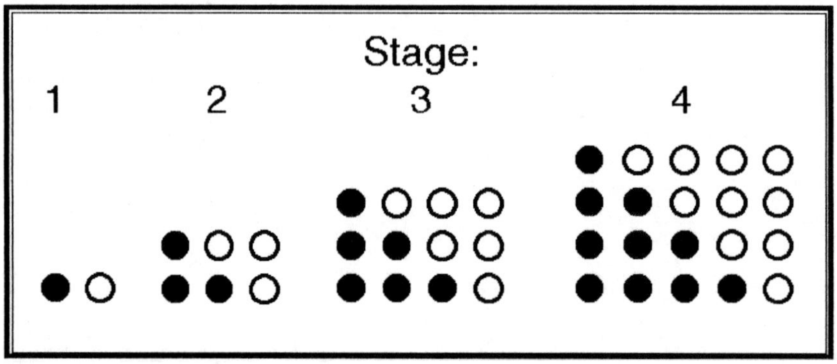

1. Term Number	**1**	**2**	**3**	**4**	**5**	**6**	**7**	**8**	***n***
2. Value	1	3	6	10	15	21	28	36	
3. Twice the Value	2	6	12	20	30	42	56	72	
4. Factors for Row 3	**1** • 2	**2** • 3	**3** • 4	**4** • 5	**5** • 6	**6** • 7	**7** • 8	**8** • 9	***n***(*n* + 1)

Since the formula represents twice the actual triangular numbers, you must divide by 2 to find the formula for the triangular numbers. The formula is: $y = \frac{1}{2}\, n(n + 1)$ or $y = .5\, n(n + 1)$. Compare this formula to the function in **Y1**.

1. Press [LIST]. Move the cursor to the top of **L3**.

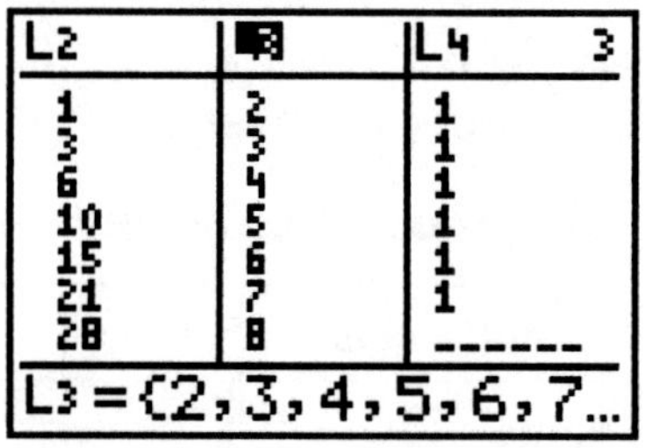

2. Press [CLEAR] [ENTER] to clear **L3.** Press [▲] to highlight **L3** again.

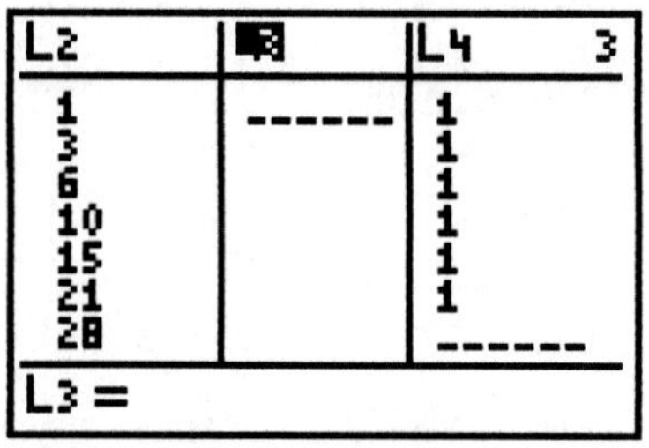

3. Enter the formula for the sequence, .5n(n+1), except substitute **L1** for *n*. Press [.] **5** [2nd] [STAT] **1:L1** [×] [(] [2nd] [STAT] **1:L1** [+] **1** [)].

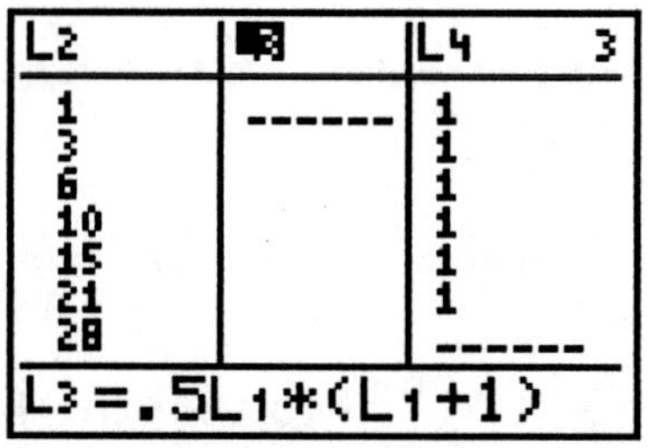

4. Press [ENTER] to calculate the values. Compare the values found in **L2** and **L3**.

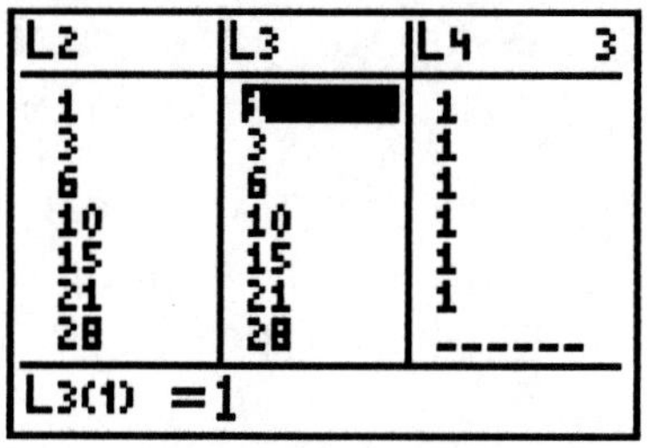

There are two types of formulas used to represent sequences. The type you have just found is an example of an explicit formula. (You will investigate the other type later in this activity.) An *explicit formula* is a formula that tells how to find the value of any term of a sequence without finding all the previous terms.

Complete the **Entering the data, Graphing the data,** and **Analyzing the data** sections using square numbers.

Answer questions 5 - 7 on the **Data Collection and Analysis** page.

Finding a different formula

As mentioned before, there are two different types of formulas used to represent sequences. The type you are going to investigate is called a recursive formula. A recursive formula is a formula that tells how to find a term using the previous term in the sequence. You will now investigate a recursive formula for triangular numbers.

Study the pattern listed below to see how the recursive formula is developed.

1. Term Number	**1**	**2**	**3**	**4**	**5**	**6**	**7**	**8**	***n***
2. Value	1	3	6	10	15	21	28	36	
3. Formula	**1**+ 0	**2** + 1	**3** + 3	**4** + 6	**5** + 10	**6** + 15	**7** + 21	**8** + 28	**?**

To develop the next term in the sequence, you must add the term number to the value of the previous term. Use **Next** to indicate the next value and **Previous** to indicate the previous value.

Next Value = Previous Value + Term Number

For example: Value of term 4 = Value of term 3 + Term Number

10 = 6 + 4

Note: *6 is the value of the* 3rd *term and 4 is the term number.*

Using the TI-73 to test the formula

1. Generate the terms of the sequence. Press [2nd] [QUIT] to go to the Home screen. Press [CLEAR] to clear the Home screen.

2. Press [2nd] [TEXT]. Press [▼] [▼] [◄] [◄] [◄] [◄] [◄] to highlight the opening brace, "{". Press [ENTER] to type "{".

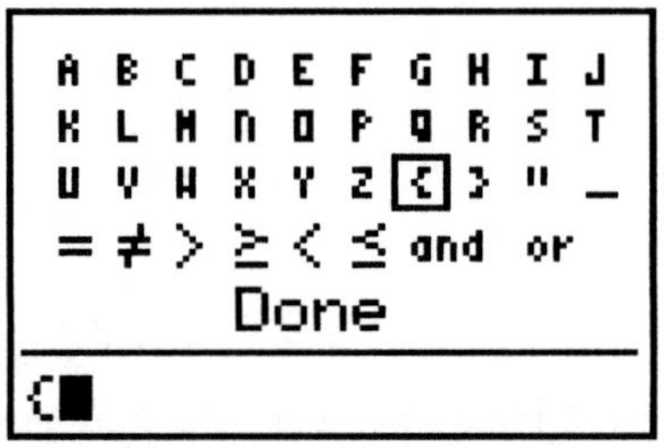

3. Press **1** [,] **1**. These are the term number and the value of the first term, respectively.

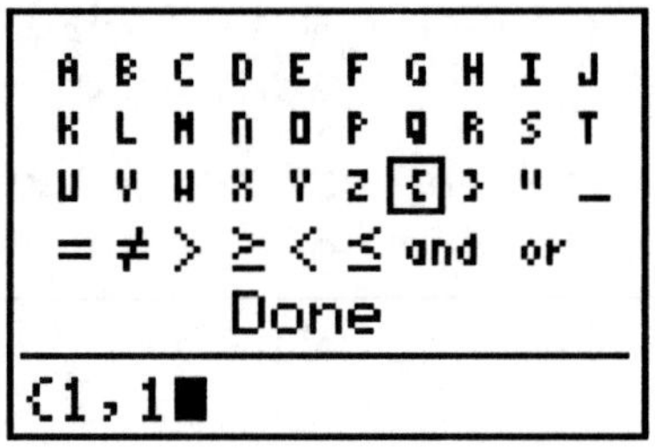

4. Close the expression by pressing ▶ ENTER to type the closing brace, "}".

```
A B C D E F G H I J
K L M N O P Q R S T
U V W X Y Z { } " _
= ≠ > ≥ < ≤ and or
        Done
─────────────────────
{1,1}
```

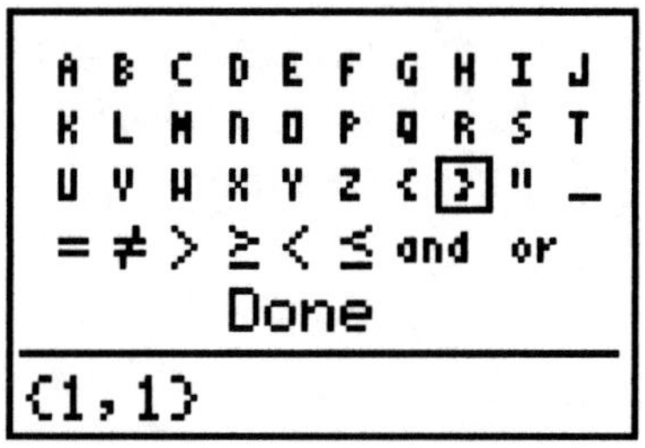

5. Press ▼ ▼ ENTER to exit the Text editor.

```
{1,1}
```

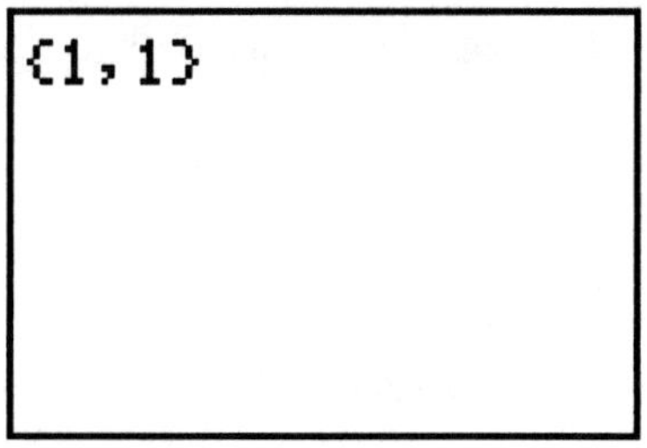

6. Press ENTER to store these values on the TI-73. We will refer to the term number as **Ans(1)**, and to the value of the term as **Ans(2)**.

```
{1,1}
            {1  1}
```

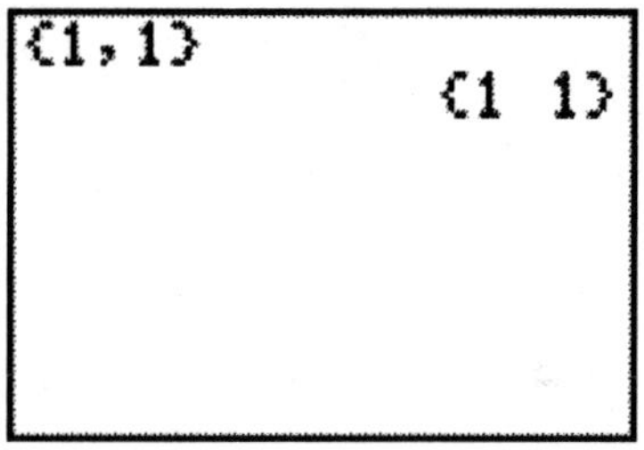

7. Press 2nd [TEXT]. Press ▼ ▼ ◀ ◀ ◀ ◀ ◀ ◀ to highlight the opening brace, "{". Press ENTER to type "{".

```
A B C D E F G H I J
K L M N O P Q R S T
U V W X Y Z { } " _
= ≠ > ≥ < ≤ and or
        Done
─────────────────────
{■
```

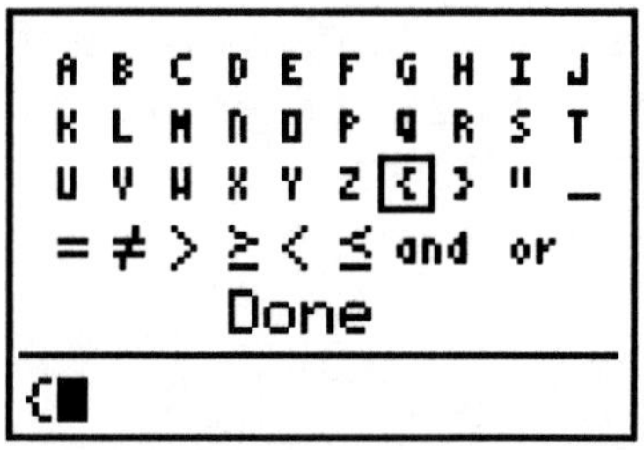

8. Press ▼ ▼ ENTER to exit the Text editor.

```
{1,1}
            {1  1}
{■
```

9. Press 2nd [ANS] (1) + 1 , 2nd [ANS] (2) + 2nd [ANS] (1) + 1.

```
{1,1}
            {1  1}
{Ans(1)+1,Ans(2)
+Ans(1)+1■
```

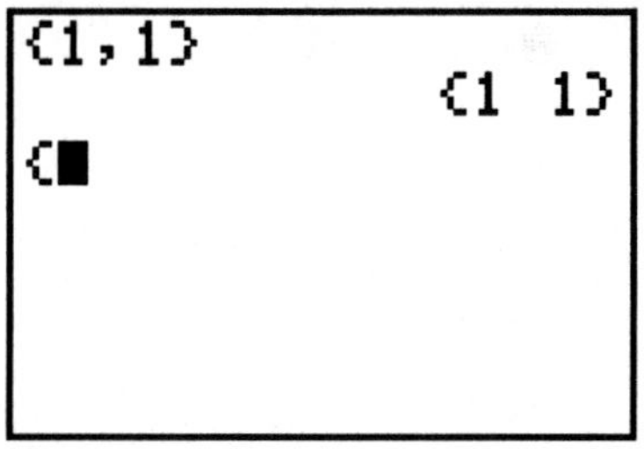

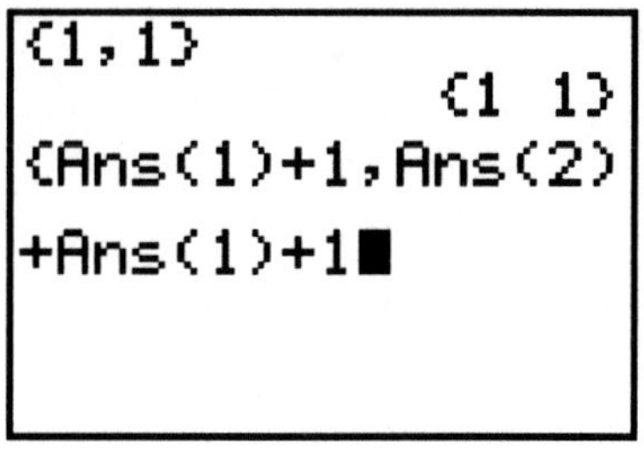

10. Press [2nd] [TEXT]. Press [▼] [▼] [◀] [◀] [◀] to highlight the closing brace, "}". Press [ENTER] to type "}".

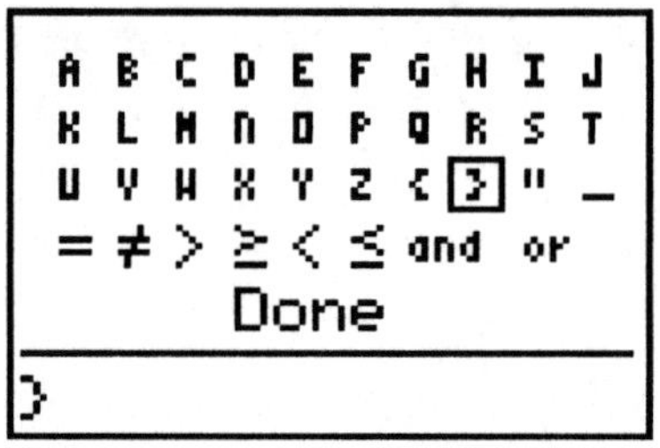

11. Press [▼] [▼] [ENTER] to exit the Text editor.

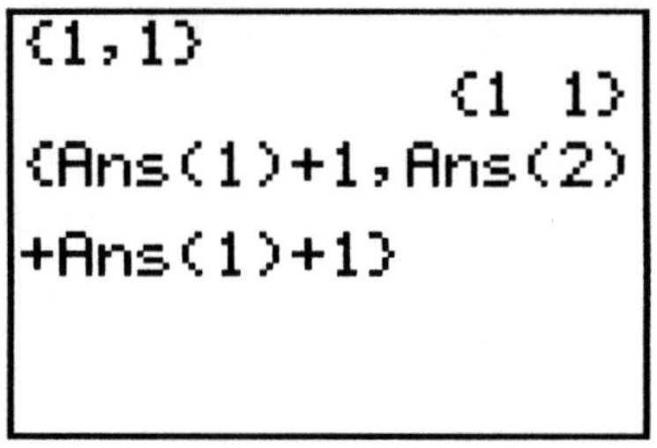

12. Press [ENTER] to generate the second term in the sequence.

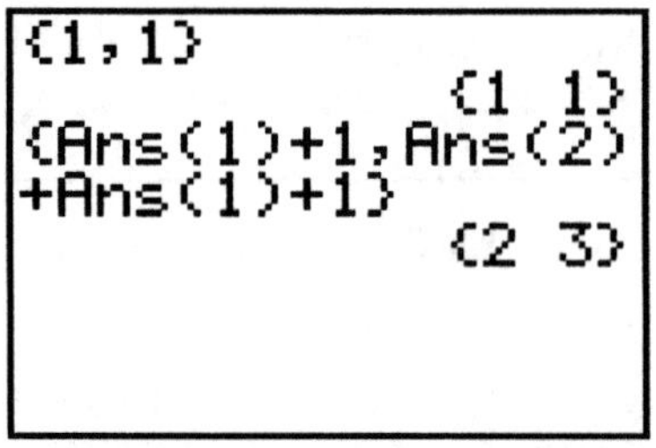

13. Continue to press [ENTER] to generate successive terms in the sequence.

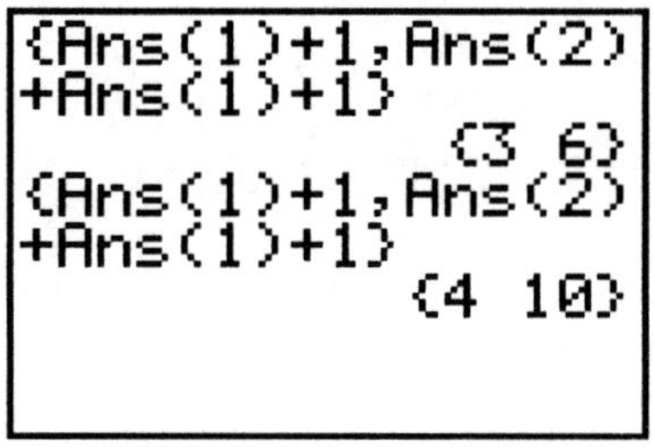

Answer questions 8 – 10 on the **Data Collection and Analysis** page.

Data Collection and Analysis

Name ______________________

Date ______________________

Activity 11: Get Your Numbers in Shape

Collecting the data

Record your data for triangular numbers in the table below.

Stage Number	1	2	3	4	5	6	7	8
Value	1	3	6	10				

Analyzing the data

Use your equation from number 8 in the **Analyzing the data: Finding a trend line** section to answer questions 1 through 4.

1. Find the 20th *triangular number.* ______________________

2. Find the 100th *triangular number.* ______________________

3. What is the stage for a *triangular number* that equals 120? ____________

4. What is the stage for a *triangular number* that equals 496? ____________

5. Find an explicit formula for the *square numbers.* ______________

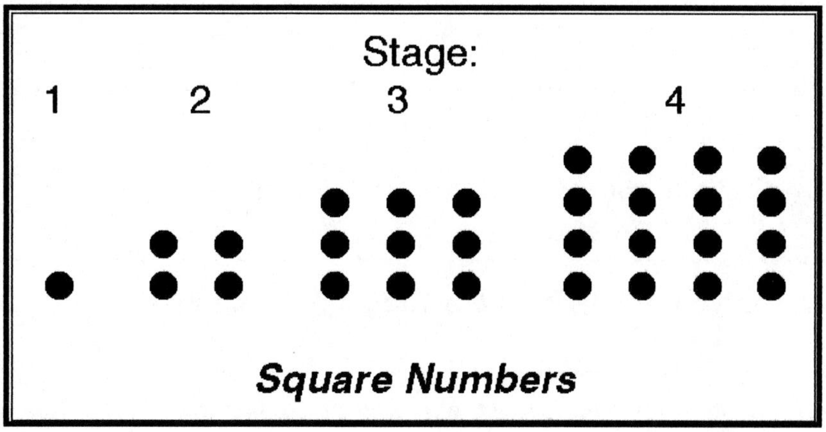

6. Find an explicit formula for the pentagonal numbers shown below.

____

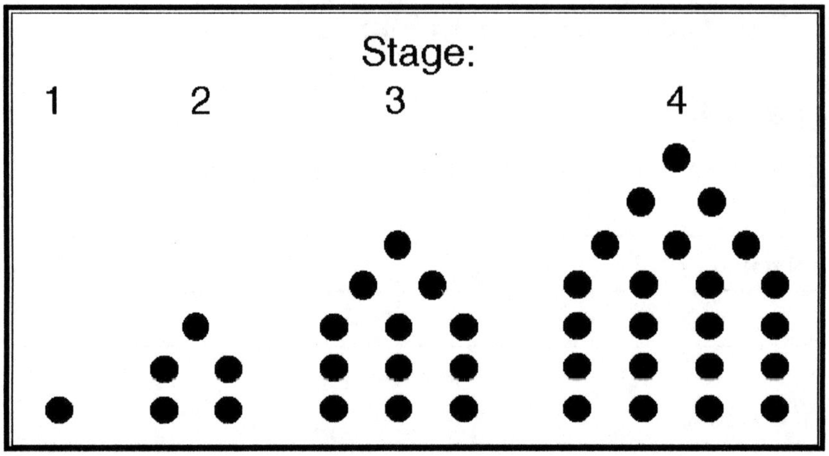

7. You go to a back-to-school party. Thirty of your friends are at the party. After the long summer break, you are happy to see all of your friends. You begin shaking hands with each person at the party. In fact, each person at the party shakes hands with everyone else at the party. How many handshakes are there altogether?

____

8. Did the values for the triangular numbers sequency match the values you found using the recursive formula? Explain.

____

____

____

9. Find a recursive formula for the square numbers.

____

10. Find a recursive formula for the pentagonal numbers.

____

***Note**: For questions 9 and 10, follow the steps in the **Using the TI-73 to test the formula section** to check your work.*

Teacher Notes

Activity 11

Get Your Numbers in Shape

Objectives

- To use technology to explore patterns
- To use inductive reasoning to make conjectures about patterns
- To use technology to produce a sequence
- To find the *x* value of a function, given the *y* value
- To find the *y* value of a function, given the *x* value
- To find a linear or a quadratic equation for a given pattern

Materials

- TI-73 graphing device
- Small cubes or candy

Preparation

For questions 6 and 10, give the students the following hint.
Hint: Look at the dots. What two figurate patterns form the pentagonal pattern? Look at the formulas for these two patterns.

Management

For question number 7, act out the situation starting with a smaller group of students. Develop a table to keep track of the results. Gradually increase the number of students until there are 20 students. To help the students visualize this, have students draw a polygon, where the number of vertices of the polygon represent the number of students.

Answers to Data Collection and Analysis

Collecting the data

Sample data:

Term Number	1	2	3	4	5	6	7	8
Value	1	3	6	10	15	21	28	36

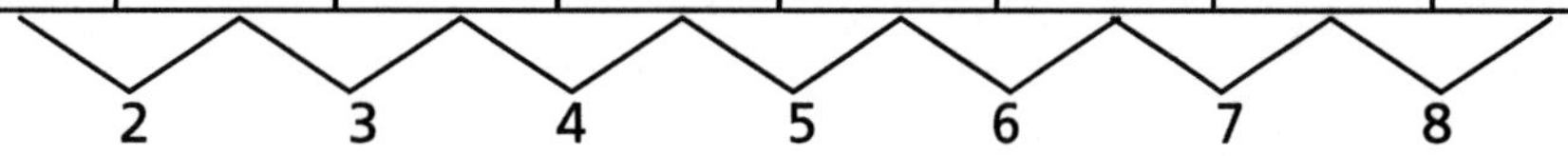

Analyzing the data

Use your equation from number 8 in the **Analyzing the data: Finding a trend line** section to answer questions 1 through 4.

1. Find the 20th *triangular number.*

   210

2. Find the 100th *triangular number.*

   5050

3. What is the stage for a *triangular number* that equals 120?

   15

4. What is the stage for a *triangular number* that equals 496?

   31

5. Find an explicit formula for the square numbers.

   $y = x^2$

6. Find an explicit formula for the pentagonal numbers.

   $Y = 1.5x^2 - 0.5x$

7. You go to a back-to-school party. Thirty of your friends are at the party. After the long summer break, you are happy to see all of your friends. You begin shaking hands with each person at the party. In fact, each person at the party shakes hands with everyone else at the party. How many handshakes are there altogether? *formula:* $y = \frac{1}{2} n(n - 1)$

   For 31 people there are 465 handshakes.

8. Did the values for the triangular numbers sequence match the values you found using the recursive formula? Explain.

   Yes. Both formulas give the same sequence of numbers.

9. Find a recursive formula for the *square numbers.*

   For any term, n = previous term + (2n + 1)

10. Find a recursive formula for the pentagonal numbers shown below. $n(n + 1)$

    For any term, n = previous term + (3n - 2)

EXPLORATIONS

Activity 12

Murder in the First Degree — The Death of Mr. Spud

Objectives

- To model the process of cooling
- To use a cooling curve to simulate a forensic scenario to predict the time of death
- To use technology to find an exponential plot

Materials

- TI-73 graphing device
- CBL 2™ data collection device (optional)
- Small potato
- Pot with boiling water
- Containers of ice water
- Extra ice
- Celsius thermometer
- Terperature probe (optional)
- Stop watch

Introduction

In a murder investigation, a forensic expert may be called in to determine the time of death. Such determinations may involve examining the contents of the victim's stomach or inspecting decomposing insects on the body. One interesting approach is to examine the temperature of the body. Human body temperature is approximately 37 degrees Celsius. Immediately after a person dies, the body temperature begins to drop. By determining how far the temperature has dropped, you may be able to arrive at an accurate measure of the time of death. This information could play an important role in either the prosecution or defense of an alleged criminal.

Problem

A potato is placed in boiling water. After removing the potato from the boiling water, it begins to cool, just as a human body cools after death. By examining the temperature of a potato, determine the time of death (removal from the boiling water).

Collecting the data

Forensic experts measure the temperature drop in corpses in order to establish *standard curves* under controlled conditions. When a person is found dead and foul play is suspected, the forensic expert measures the temperature of the body. The forensic pathologist can approximate the time of death by determining where the temperature is on the standard curve.

You will simulate the drop in a person's body temperature at the time of death. Your teacher placed a potato in a pot of boiling water and allowed it to stand for 15 minutes. Removal of the potato from the boiling water will simulate the "death" of the potato. When the potato is removed from the pot, its temperature will drop toward the temperature of its surroundings, just as the temperature of a body drops following death. You will plot this data to establish a standard curve.

(Your teacher may ask you to collect data using a CBL 2™ data collection device with a temperature probe or may provide you with data for the standard curve.)

1. Get a potato with a thermometer from the teacher that has been in boiling water.
2. Immediately place the potato and thermometer into ice water.
3. Record the temperature and enter the reading as time 0 in the table on the **Data Collection and Analysis** page.
4. Record the temperature every five minutes for 30 minutes.

Setting up the TI-73

Before starting your data collection, make sure that the TI-73 has the STAT PLOTS turned OFF, Y= functions turned OFF or cleared, the MODE and FORMAT set to their defaults, and the lists cleared. See the Appendix for a detailed description of the general setup steps.

Entering the data in the TI-73

1. Press [LIST]. The data lists are displayed.

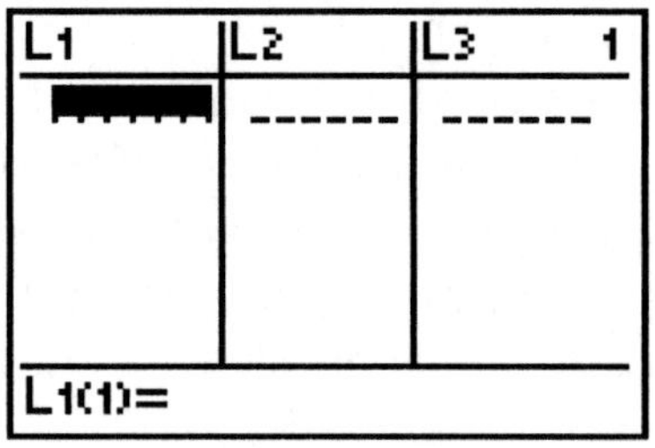

2. Enter the time in **L1**.
3. Enter the temperature in **L2**. (Make sure that the pairs of time and temperature match in each column.)

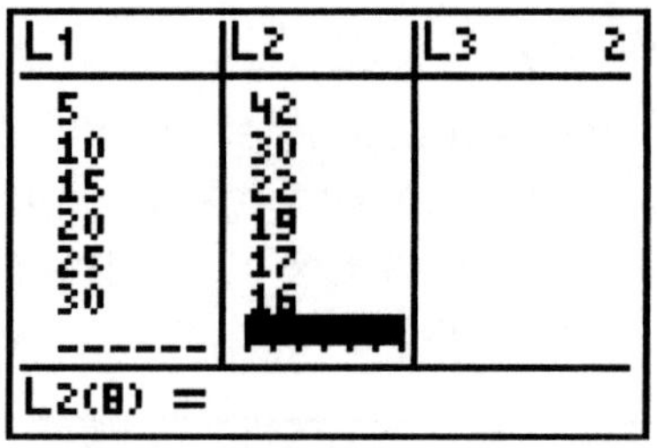

L1 L2 L3 2
5 10 15 20 25 30
42 30 22 19 17 16
L2(8) =

Setting up the window

1. Press [WINDOW] to set up the proper scale for the axes.
2. Set the **Xmin** value by identifying the minimum value in **L1**. Choose a number that is less than the minimum.

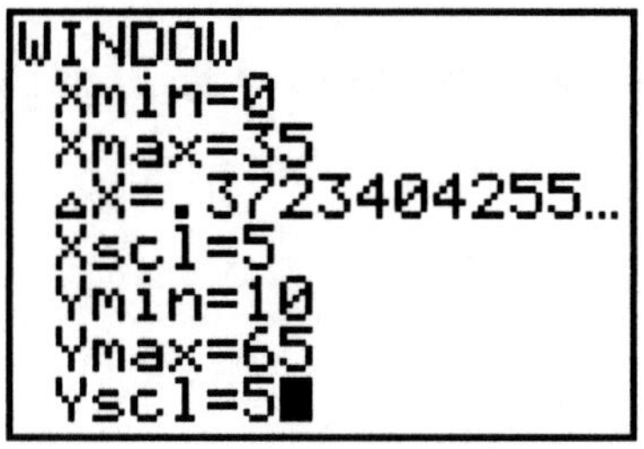

3. Set the **Xmax** value by identifying the maximum value in each list. Choose a number that is greater than the maximum. **Do Not Change ΔX Value.** Set the **Xscl** to **5**.
4. Set the **Ymin** value by identifying the minimum value in **L3**. Choose a number that is less than the minimum.
5. Set the **Ymax** value by identifying the maximum value in **L3**. Choose a number which is greater than the maximum. Set the **Yscl** to **5**.

Graphing the data: Setting up a scatter plot

In order to analyze the data, you will need to set up a scatter plot and model the data using an exponential model. You will then use the exponential model as the standard curve to predict the time of death of the potato.

1. Press [2nd] [PLOT]. Select **1:Plot1** by pressing **1** or [ENTER].

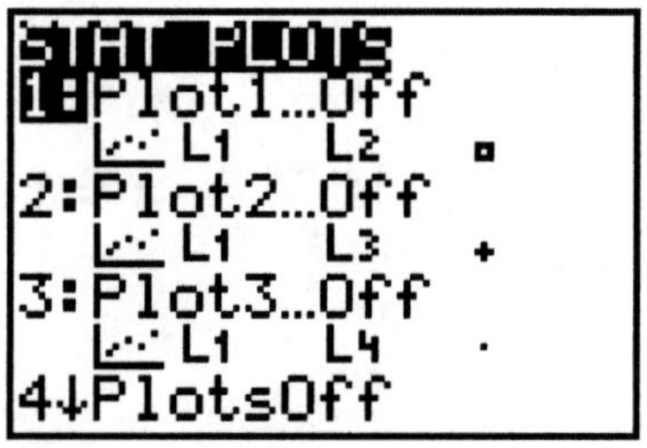

2. Set up the plot as shown by pressing [ENTER] [▼] [ENTER] [▼] [2nd] [STAT] **1:L1** [▼] [2nd] [STAT] **2:L2** [▼] [ENTER].

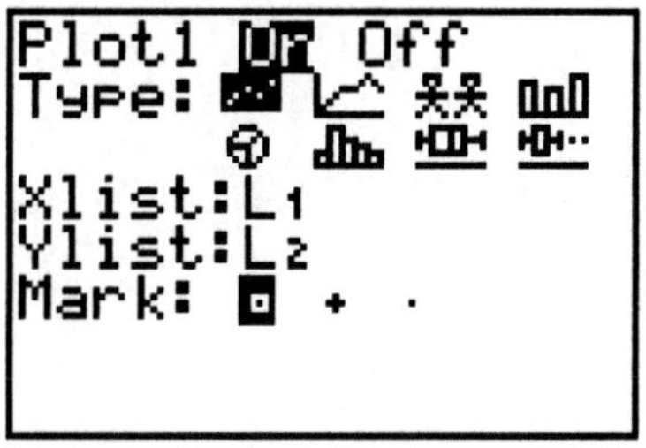

3. Press [GRAPH] to see the plot.

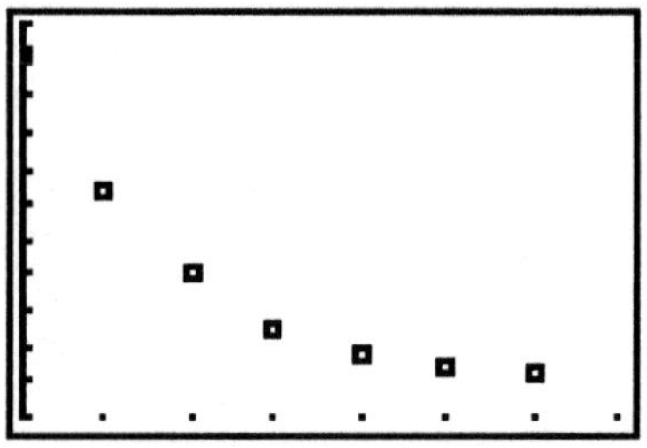

It is necessary to determine an appropriate regression model for this data. Does the plot appear to be linear? If not, how does the slope change?

Analyzing the data

Finding an exponential regression

1. Press [2nd] [STAT] [◄] to move the cursor to the **CALC** menu.

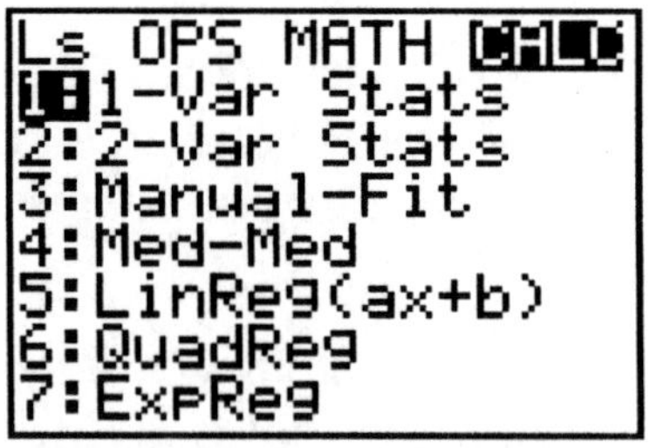

2. Select **7:ExpReg** by pressing **7**.

3. Press [2nd] [STAT] **1:L1** [,] [2nd] [STAT] **2:L2** [,].

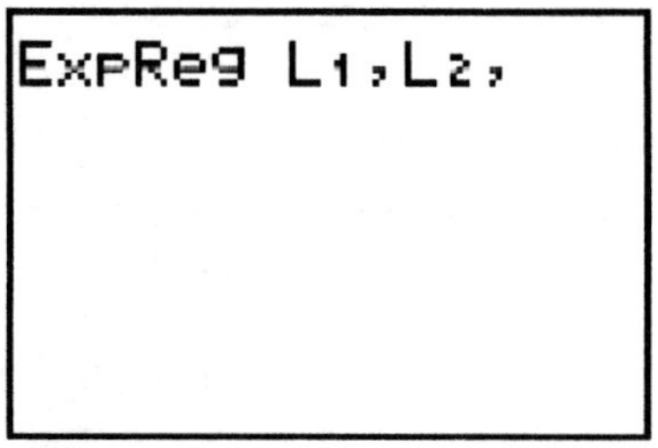

4. Press [2nd] [VARS]. Select **2:YVars** by pressing **2**.

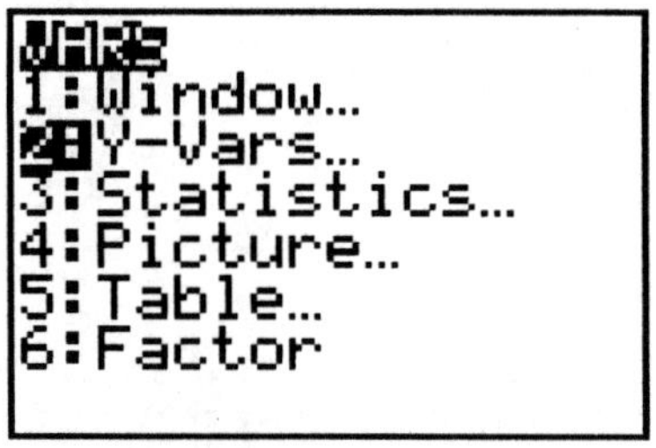

5. Select **1:Y1** by pressing **1** or [ENTER].

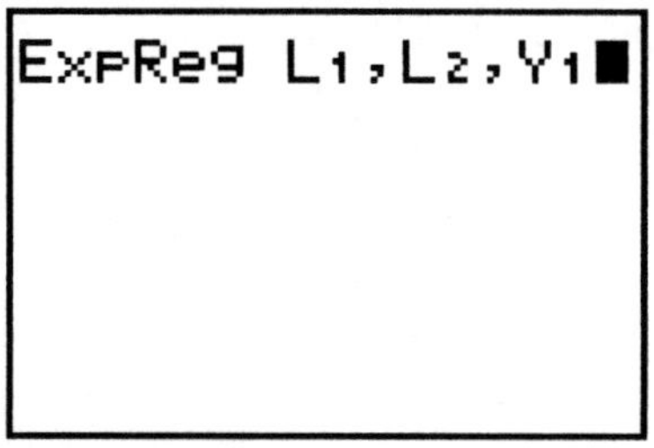

6. Press [ENTER] to calculate the exponential regression. The function is pasted in **Y1**.

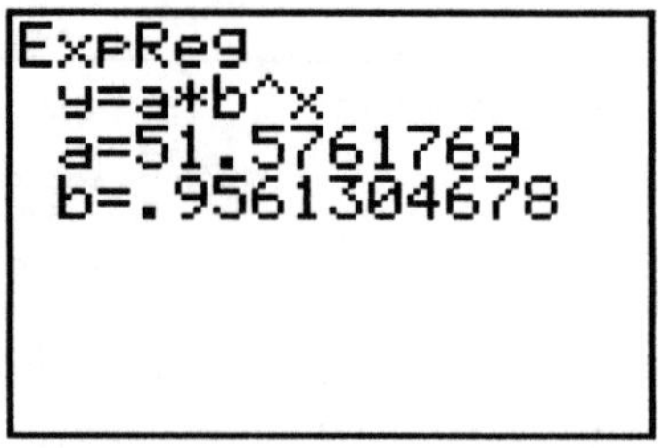

7. Press [GRAPH] to see the exponential model.

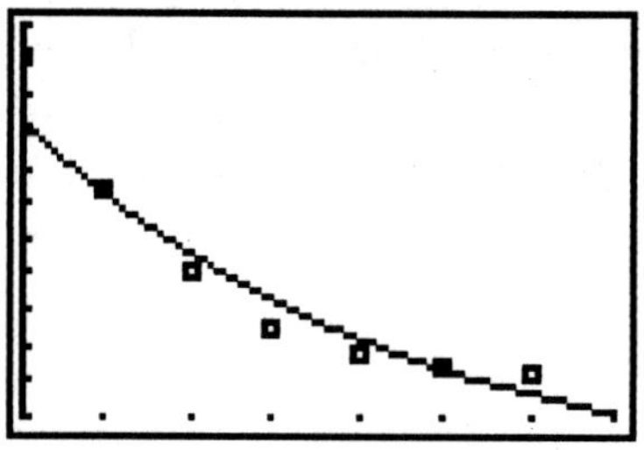

Determining the time of death

You will be given a potato whose time of death (removal from the boiling water) is unknown. This partially cooled potato will simulate a murder victim. Assume that the potato was murdered outdoors during the winter.

You will determine the time of death based on the temperature of the potato using the standard curve you developed.

1. Your teacher will give you a second potato. Record the temperature on the **Data Collection and Analysis** page.

2. Press [Y=] and move the cursor to **Y2**. Enter the temperature of the potato. In this example, 28 degrees C was used.

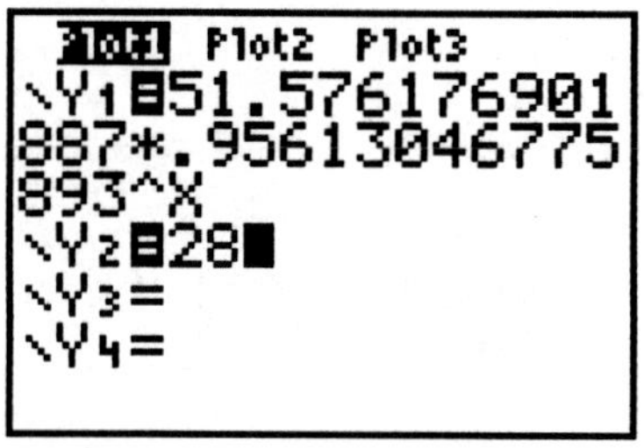

3. Press [GRAPH] to see the intersection of the two lines. The *x* value of the point where the two functions intersect is the time (in minutes) that the potato has been "dead."

   The **Table** function of the TI-73 will be used to determine the coordinates of the point of intersection.

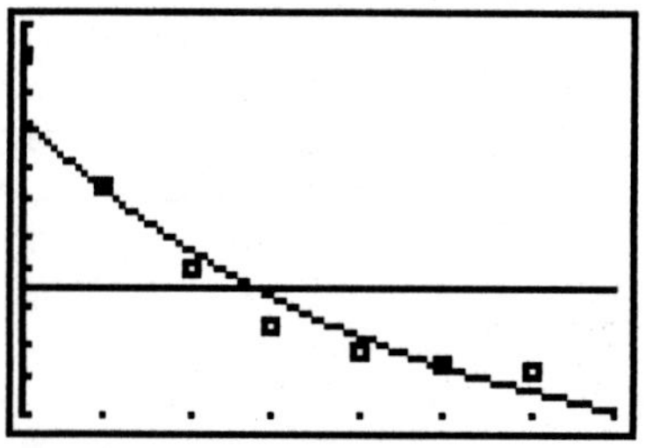

4. Press [2nd] [TBLSET]. Type in the lowest value in **L1** (0 in the example). Press [▼] **5** to set the **ΔTbl** value.

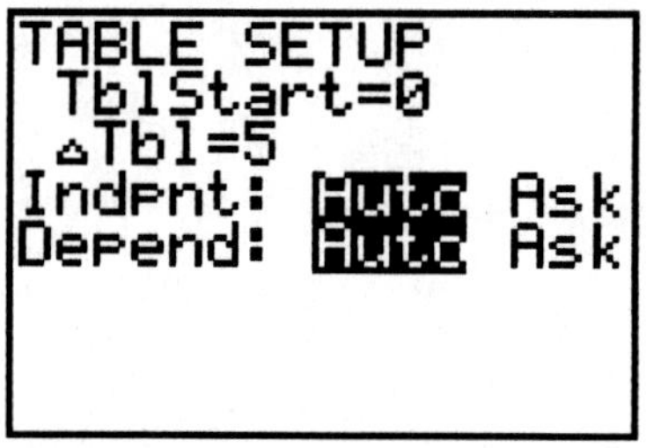

5. Press [2nd] [TABLE]. If necessary, use [▲] and [▼] to scroll the table.

   Note: *For this example, in the **Y1** column, 28 degrees Celsius falls between 32.932 and 26.315 which corresponds to 10 and 15 minutes. Based on that information, the table will be readjusted.*

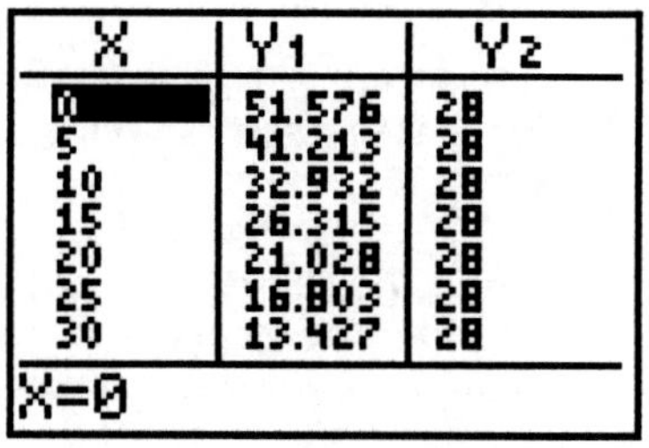

X	Y1	Y2
0	51.576	28
5	41.213	28
10	32.932	28
15	26.315	28
20	21.028	28
25	16.803	28
30	13.427	28

X=0

6. Press [2nd] [TBLSET]. Enter the results from Step 6. Press [▼] **1** to set the **ΔTbl** value.

7. Press [2nd] [TABLE]. If necessary, use [▲] and [▼] to scroll the table.

   Note: *For this example, in the* ***Y1*** *column, 28 degrees Celsius falls between 28.785 and 27.523 which corresponds to 13 and 14 minutes. Based on that information, the table will be readjusted.*

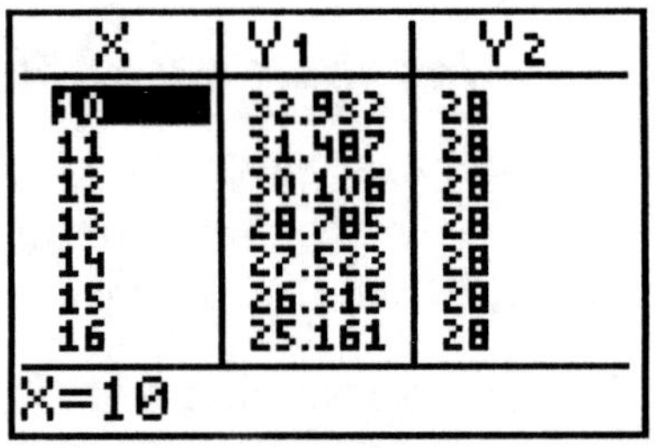

8. Press [2nd] [TBLSET]. Enter the results from Step 8. Press [▼] **0.1** to set the **ΔTbl** value.

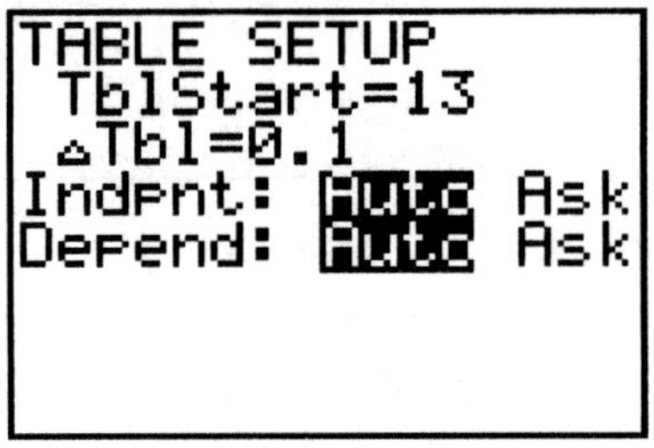

9. Press [2nd] [TABLE]. If necessary, use [▲] and [▼] to scroll the table.

   Note: *For this example, in the* ***Y1*** *column, 28 degrees Celsius falls between 28.021 and 27.896 which corresponds to 13.6 and 13.7 minutes. Time measurements were made to the nearest minute. Therefore, we will report our result using the same level of precision. From the table, the time falls between 13.6 and 13.7. Rounding to the nearest minute, the coordinates of the point of interestion are (14, 28).*

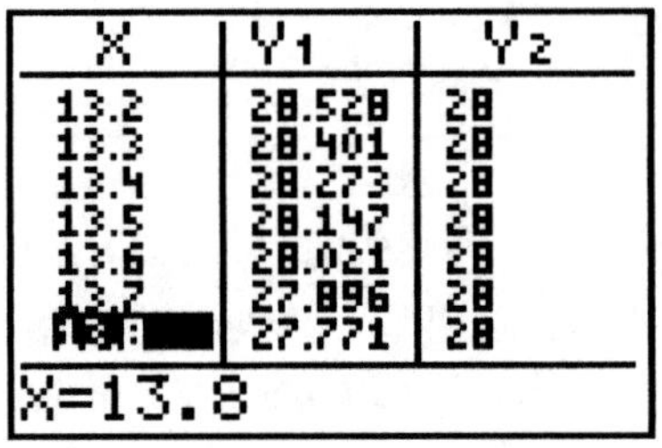

To verify this graphically, we will use the **DRAW** function.

10. Press [DRAW]. Select **4:Vertical** by pressing **4**.

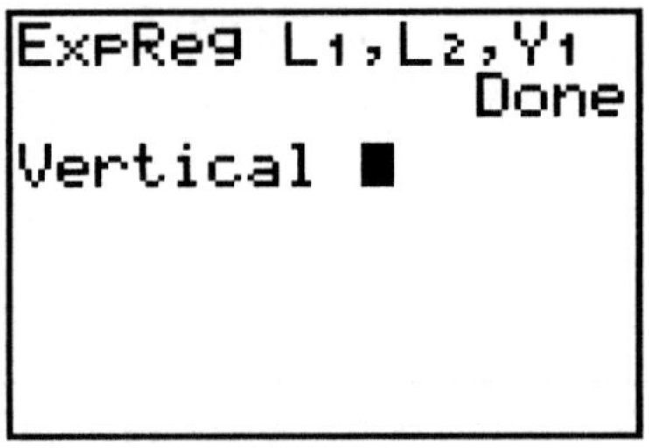

11. Type the results from Step 10, and press [ENTER].

    Note: *The coordinates of the point on the exponential model where all of the curves intersect is defined, for this example, by the vertical drawn at x=14 and the horizontal at y=28.*

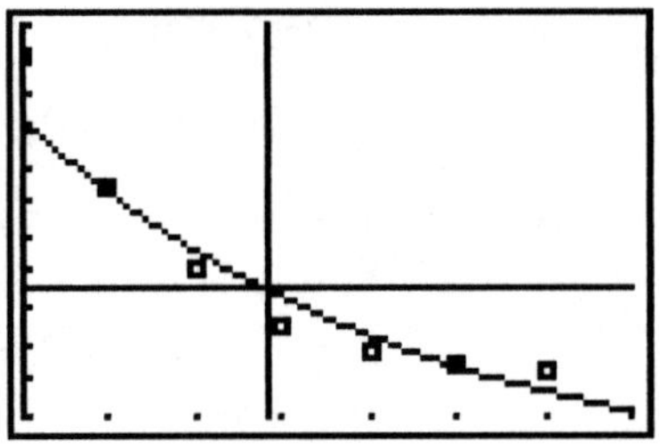

12. The coordinates can be added onto the screen by pressing [DRAW] **7:Text**, moving the cursor near the intersect point, and typing your results.

__Note:__ Text appears to the below and to the right of the cursor.

Record the point of intersection and answer questions 1 through 5 on the **Data Collection and Analysis** page.

Data Collection and Analysis

Name ______________________

Date ______________________

Activity 12: Murder in the First Degree — The Death of Mr. Spud

Collecting the data

Time (minutes)	0	5	10	15	20	25	30
Temperature (°C)							

Temperature of the potato: ______________________

Time of death of the potato: ______________________

Analyzing the data

1. Describe the shape of the temperature versus time plot.

____

____

2. What temperature does the plot appear to be approaching? What does that temperature correspond to in the situation being studied?

____

3. Explain why the plot is not linear. (**Hint:** If the plot was linear, it would mean that the temperature dropped at a constant rate. Why would such a drop not be expected?)

____

____

4. In what way(s) is this simulation not consistent with a real forensic study of the drop in body temperature following death?

____

____

5. Which model - exponential or linear - would be the better model to describe the following situations? Explain your answer in each case.

a. A car slowing down by 1/3 of the speed it was going the previous second.

____

____

b. A car slowing down 5 mph each second.

____

____

Extensions

1. In this activity, you determined the time of death by the drop in temperature. A similar problem that has a similar solution (conceptually) is to date the time of death when an organism died *thousands* of years ago. To solve this problem, scientists use a technique called *radioactive dating*. It is based on the decay of a substance called carbon-14 (C-14).

Most carbon in a living organism is called carbon-12 (C-12) and it does not decay. There is a small amount of C-14 in all living tissue. Once the organism dies, however, it no longer incorporates this substance in its body. The C-14 begins to decay. By examining how much C-14 is present (the ratio of C-12: C-14), you can determine the time since death. Carbon-14 has a half-life of about 5,600 years (actual half-life is 5,780 years). Instead of examining the drop in temperature of a corpse, radioactive dating is based on the decay of C-14 over thousands of years.

Consider the problem of dating an animal's remains. Assume that by analyzing the amount of carbon in its remains, you believe that it originally had 1000 grams of C-14, but now has only 100 grams.

Determine the amount of C-14 that remains after each 5,600-year interval, starting with 1000 grams of the substance.

Time (years)	5600	11200	16800	22400	28000	33600	39200
C-14 (Grams)	1000						

Amount of C-14 in Recovered Organism: 100 Grams

Years Since Organism Died: ____________________ yrs

2. Use two temperature probes interfaced to a CBL 2™ (Calculator Based Laboratory 2) to monitor the drop in water temperature of two "animals": one with insulating "fur" and the other, a "naked" animal. This situation can be simulated by using 2 glass or plastic beakers. Simulate fur by taping several layers of paper toweling around the beaker. For the "naked" animal, do not put any paper toweling around the beaker. Place equal volumes of water (approximately 75 degrees Celsius) into each of the beakers. Place the temperature probes into the beakers and take a reading every minute for 10 minutes.

Analyze and compare the drop in temperature of the two "animals". Model the temperature drop similarly to the temperature drop in the potatos.

Teacher Notes

Activity 12

Murder in the First Degree — The Death of Mr. Spud

Objectives

- To model the process of cooling
- To use a cooling curve to simulate a forensic scenario to predict the time of death
- To use technology to find an exponential plot

Materials

- TI-73 graphing device
- CBL 2™ data collection device (optional)
- Small potato
- Pot with boiling water
- Containers of ice water
- Extra ice
- Celsius thermometer
- Temperature probe (optional)
- Stop watch

Preparation

- The temperature of a corpse drops after death, assuming the environmental temperature is lower than the body temperature. When the body begins to putrefy, the temperature goes up a bit then gradually cools to match the surrounding temperature. If the body temperature has already reached that of the environment, other techniques must be used to pinpoint the time of death.
- Cooling curves do not exactly follow the simple exponential model built into the TI-73. Accurate models are more complex and their use would not be appropriate in an introductory Algebra class. A more accurate model would be:

  $T_t = C + (T_o - C)\, e^{-kt}$ where T_t is the temperature of the body at time t, T_0 is the initial temperature of the body, C is the environmental temperature, and k is the cooling rate constant. The assumption is that the environmental temperature remains constant, which is often not the case.

- Use small to medium size potatoes. Different size potatoes give different standard curves, so try to use potatoes that are of similar size and shape.
- Use the point of a pencil to make a hole in each potato prior to putting them into boiling water. Use the blunt end (eraser) of the pencil to enlarge each hole so that it can accommodate a thermometer. The tip of the thermometer should reach the middle of the potato.

- It is sometimes difficult to know how far to insert the thermometer in the potato. Hold the thermometer next to the potato to see how far to insert it. You could mark the thermometer with a strip of masking tape to identify the point where the thermometer was inserted.
- As the potato cools in the ice water, be sure the students add ice to the bath so that all the ice does not melt.
- It is understood that you may not want to spend class time collecting data for the standard curve unless this activity is being coordinated between a science and math teacher. You may want to provide the data for the standard curve (see the sample data below).
- Data for the cooling curve can be obtained using a CBL 2™ data collection device. Put the temperature probe into the potato so that the tip is in the center of the potato. Immerse the potato and probe in the boiling water bath for 15 minutes. Remove the potato and collect data for 30 minutes, collecting a data point every 60 seconds.
- Give the students potatoes that were removed from the boiling water at different time intervals. Be sure to put the potatoes immediately into the ice water bath to simulate, as much as possible, the conditions modeled by the cooling curve. Use a stopwatch for each potato so you know how long they have been "dead."
- For the first **Extension** activity, it may be easier for some students to use manipulatives to better visualize radioactive decay. Here are a few suggestions:

  a. Use small candies such as M&M's® to simulate radioactive decay. Students are told to spill about 50 M&M's from a cup and remove all of the samples with the letters facing up. Ask the students to count the number of remaining M&M's. Repeat the procedure, counting the number of M&M's left after each spill.

  b. Use coins to simulate radioactive decay.
- For the second **Extension** activity, use two graduated cylinders. Measure 100 mL of the same temperature water in each. Pour the water into the beakers at the same time. Place aluminum foil over each beaker and poke the temperature probes through the foil. The tip of the probe should sit in the same position in the middle of the water in each beaker; it should **not** rest against the side of one beaker. Start the experiment when the set-ups are correct.

Answers to Data Collection and Analysis

Collecting the data

Sample data:

Time (minutes)	0	5	10	15	20	25	30
Temperature. (°C)	61	42	30	22	19	17	16

Analyzing the data

1. Describe the shape of the temperature versus time plot.

   The temperature drops sharply at first and then levels off.

2. What temperature does the plot appear to be approaching? What does that temperature correspond to in the situation being studied?

   The plot approaches the environmental temperature, in this case, 0 degrees Celsius.

3. Explain why the plot is not linear. (**Hint:** If the plot was linear, it would mean that the temperature is dropping at a constant rate. Why would such a drop not be expected?)

   A linear plot would mean that the rate at which the temperature drops is constant. The closer the potato gets to the environmental temperature, however, the less quickly it drops.

4. In what way(s) is this simulation not consistent with a real forensic study of the drop in body temperature following death?

   Answers may vary, but one difference is the fact that the environmental temperature does not stay the same following death.

5. Which model — exponential or linear — would be the better model to describe the following situations? Explain your answer in each case.

   a. A car slowing down by 1/3 of the speed it was going the previous second.

   Exponential

   b. A car slowing down 5 mph each second.

   Linear

EXPLORATIONS

Activity 13

Do You Have the Temperature?

Objectives

- To graphically represent and analyze climate data
- To use linear regressions to understand the relationship between temperatures as measured in the Fahrenheit and Celsius scale
- To use linear regressions to understand conversion factors
- To use technology to find a linear regression

Materials

- TI-73 graphing device
- CBL 2™ data collection device (optional)
- Climate data for different ecosystems
- Cold cup with ice water
- Hot cup with boiling water
- Rubber band
- Watch with a second hand
- Celsius thermometer and Fahrenheit thermometer
- Two temperature probes (per CBL 2) (optional)

Introduction

A man was recently listening to the radio in his New York apartment. The radio announcer was giving the temperatures of cities in other countries. The announcer said, "The temperature in London is 68 degrees." At that same moment, another man was sitting in his London apartment listening to the weather report. The radio announcer said, "The temperature in London is 20 degrees." Both announcers were right! How could this happen?

Problem

There are different units of temperature: Celsius and Fahrenheit. How can you mathematically show the relationship between these two units of temperature?

Collecting the data — Part I

You will need twelve temperature readings using both Celsius and Fahrenheit thermometers. These readings will be obtained in one of three ways.

1. Collect the readings during class.

   a. Take a Celsius and a Fahrenheit thermometer and rubber band them together so that the bulbs are next to each other.

b. Place the thermometers in a cup with boiling water and allow the temperature to stabilize. You and your student partner will each read one of the two thermometers. Record the temperature on the **Data Collection and Analysis** page.

c. Place the thermometers in a cup with ice water and start timing. Take a reading every ten seconds, recording the values on the **Data Collection and Analysis** page. You and your student partner will each read one of the two thermometers.

2. Use two temperature sensors connected to a CBL 2™ (Calculator Based Laboratory-2). Set the temperature sensor in **CH1** to take readings in Celsius and set the temperature sensor in **CH2** to take readings in Farenheit. Set the CBL 2 to take a reading every 5 seconds for 12 samples. Follow step **1b**, as directed. For step **1c**, move the sensors into the cold water. Press **2: START** and stir the water with the sensors during data collection.

Record the values in **L2** in the Temperature (°C) section and the values in **L3** in the Temperature (°F) section on the **Data Collection and Analysis** page.

3. Use the 12 average monthly Celsius and Fahrenheit temperatures provided by the teacher.

Setting up the TI-73

Before starting your data collection, make sure that the TI-73 has the STAT PLOTS turned OFF, Y= functions turned OFF or cleared, the MODE and FORMAT set to their defaults, and the lists cleared. See the Appendix for a detailed description of the general setup steps.

Entering the data in the TI-73

1. Press [LIST].

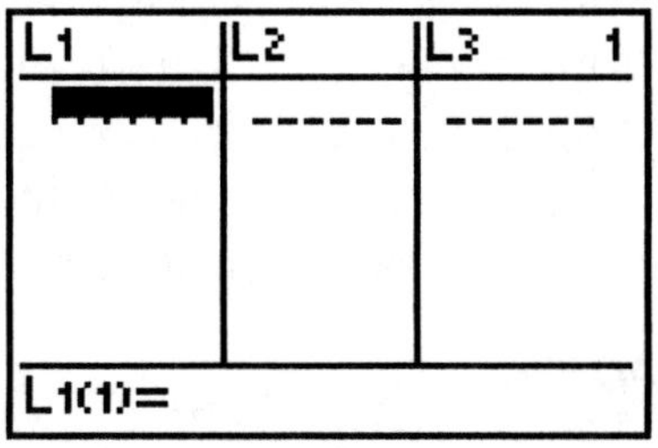

2. Enter the Celsius temperatures in **L1**.
3. Enter the Fahrenheit temperatures in **L2**. (Make sure that the pairs of Celsius and Fahrenheit temperatures match in each column.)

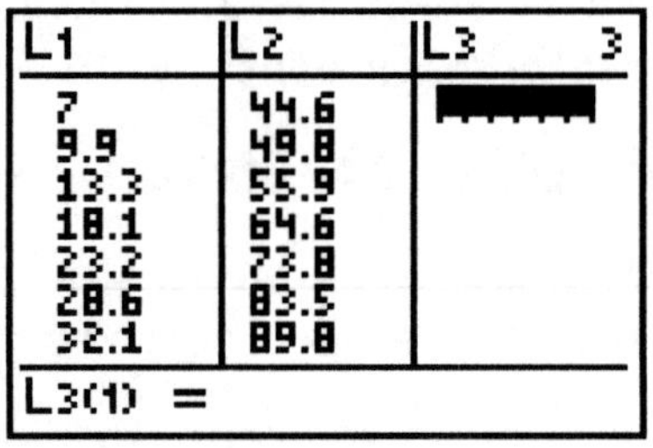

L1
L2
L3 3
7
9.9
13.3
18.1
23.2
28.6
32.1
44.6
49.8
55.9
64.6
73.8
83.5
89.8
L3(1) =

Setting up the window

1. Press [WINDOW] to set up the proper scale for the axes.

2. Set the **Xmin** value by identifying the minimum value in **L1**. Choose a number that is less than the minimum.

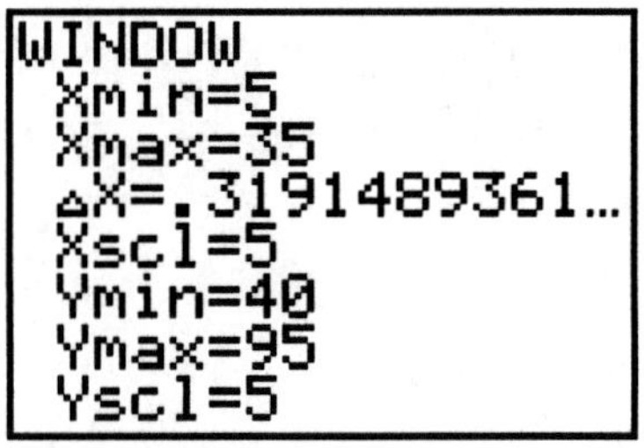

3. Set the **Xmax** value by identifying the maximum value in each list. Choose a number which is greater than the maximum. **Do Not Change the ΔX value.** Set the **Xscl** to **5**.

4. Set the **Ymin** value by identifying the minimum value in **L2**. Choose a number that is less than the minimum.

5. Set the **Ymax** value by identifying the maximum value in **L2**. Choose a number which is greater than the maximum. Set the **Yscl** to **10**.

Graphing the data: Setting up a scatter plot

In order to analyze the data, you will need to set up a scatter plot and model the data by graphing a line of best fit.

1. Press [2nd] [PLOT]. Select **1:Plot1** by pressing **1** or [ENTER].

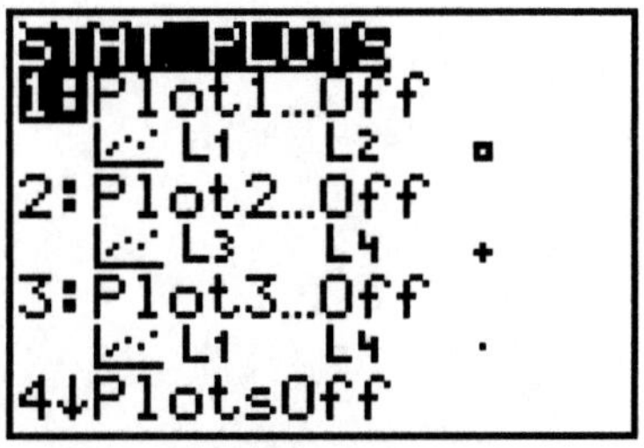

2. Set up the plot as shown by pressing [ENTER] [▼] [ENTER] [▼] [2nd] [STAT] **1:L1** [▼] [2nd] [STAT] **2:L2** [▼] [ENTER].

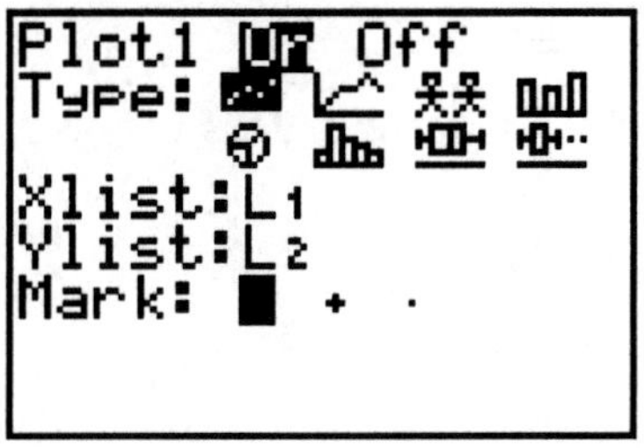

3. Press [GRAPH] to see the plot.

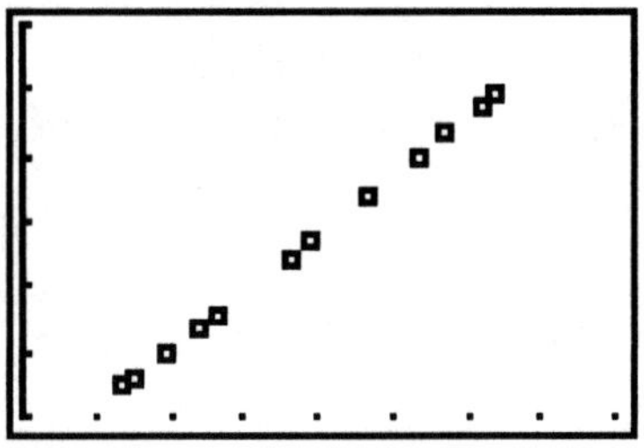

What is the appropriate regression for the plot? Does the slope change or appear constant?

Analyzing the data: Finding a linear regression

1. Press [2nd] [STAT] [◄] to move the cursor to the **CALC** menu.

    ```
    Ls OPS MATH CALC
    1▮1-Var Stats
    2:2-Var Stats
    3:Manual-Fit
    4:Med-Med
    5:LinReg(ax+b)
    6:QuadReg
    7:ExpReg
    ```

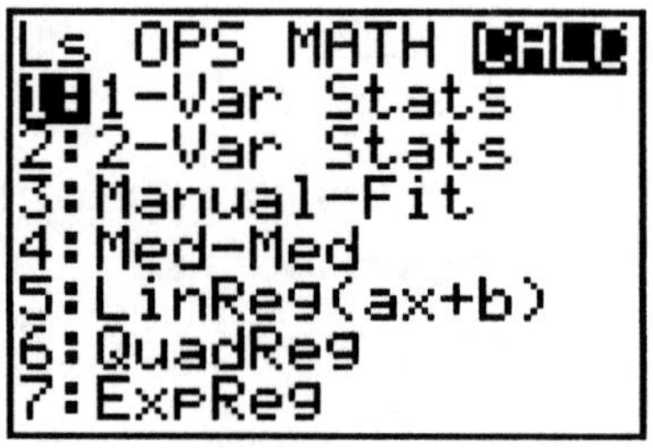

2. Select **5:LinReg(ax+b)** by pressing **5**.

    ```
    LinReg(ax+b) ▮
    ```

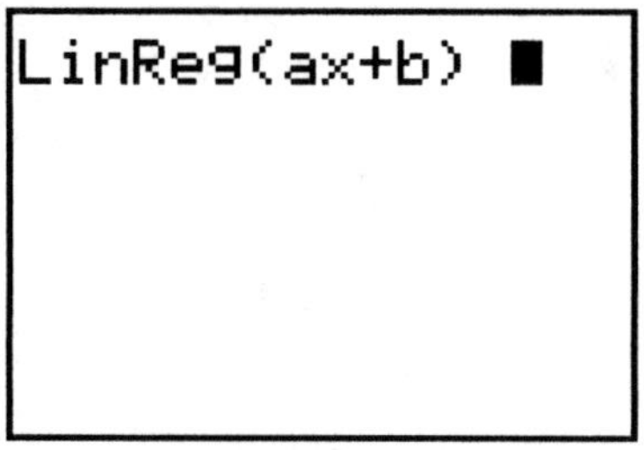

3. Press [2nd] [STAT] **1:L1** [,] [2nd] [STAT] **2:L2** [,].

    ```
    LinReg(ax+b) L₁,
    L₂,▮
    ```

4. Press [2nd] [VARS].

    ```
    VARS
    1▮Window…
    2:Y-Vars…
    3:Statistics…
    4:Picture…
    5:Table…
    6:Factor
    ```

5. Select **2:Y-Vars** by pressing **2**.

    ```
    FUNCTION
    1▮Y₁
    2:Y₂
    3:Y₃
    4:Y₄
    5:FnOn
    6:FnOff
    ```

6. Select **1:Y1** by pressing **1** or [ENTER].

    ```
    LinReg(ax+b) L₁,
    L₂,Y₁▮
    ```

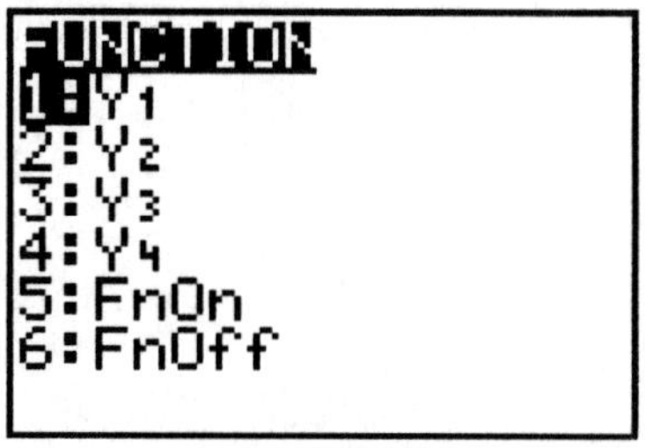

7. Press [ENTER] to calculate the linear regression. The function is pasted in **Y1**.

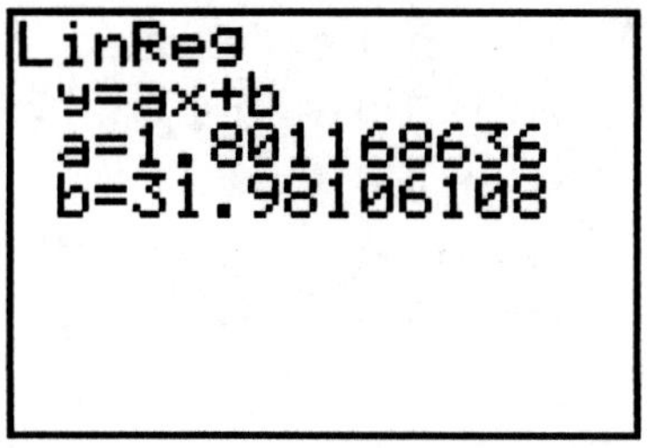

8. Press [GRAPH] to see the linear regression.

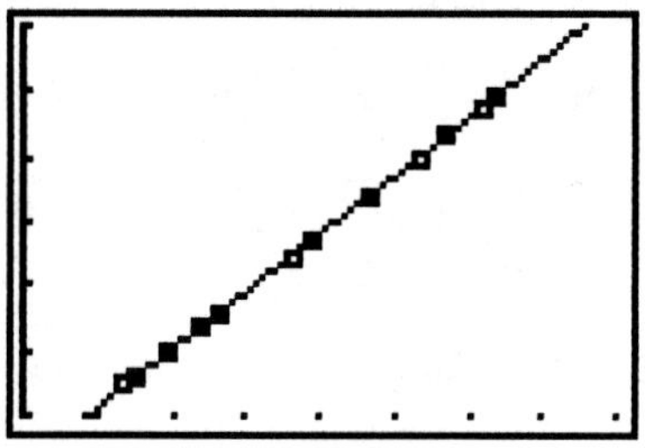

Answer Part I questions 1 through 6 on the **Data Collection and Analysis** page.

Collecting the data — Part II

Thus far, you have used the data to verify an equation for converting degrees Celsius to degrees Fahrenheit. What if you needed an equation for the opposite conversion, degrees Fahrenheit to degrees Celsius? Simply reversing the data in a **STAT PLOT** and then determining the linear regression for the data could produce this equation.

Setting up the window

1. Press [WINDOW] to set up the proper scale for the axes.
2. Set both the **Xmin** and **Ymin** values to *–50*.
3. Set the **Xmax** and **Ymax** values by identifying the maximum value in both lists. Choose a number greater than the maximum.

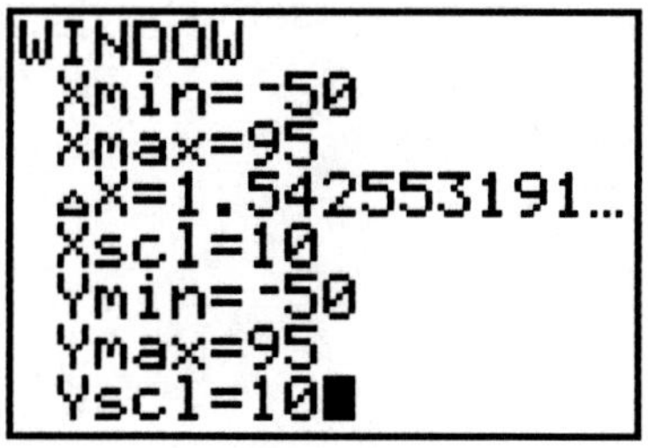

4. Set both the **Xscl** and **Yscl** values to **10**. **Do Not Change the ΔX Value**.

Graphing the data: Setting up a scatter plot

1. Press [2nd] [PLOT]. Select **2:Plot2** by pressing **2**.

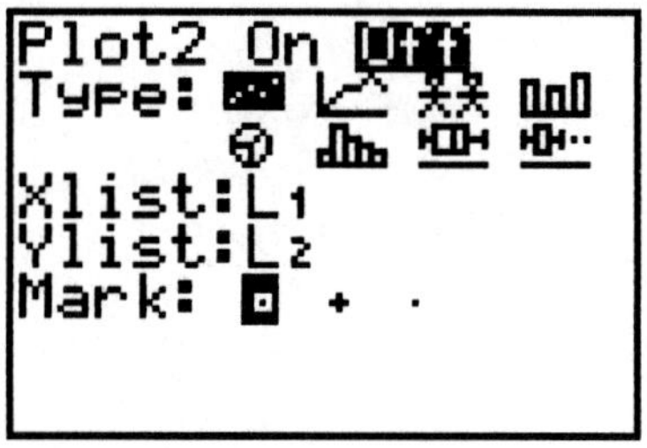

2. Set up the plot as shown by pressing [ENTER] [▼] [ENTER] [▼] [2nd] [STAT] **2:L2** [▼] [2nd] [STAT] **1:L1** [▼] [▶] [ENTER].

   Note: *This is similar to* **Plot1**, *except that* **L2** *is now the* **Xlist** *and* **L1** *is now the* **Ylist.**

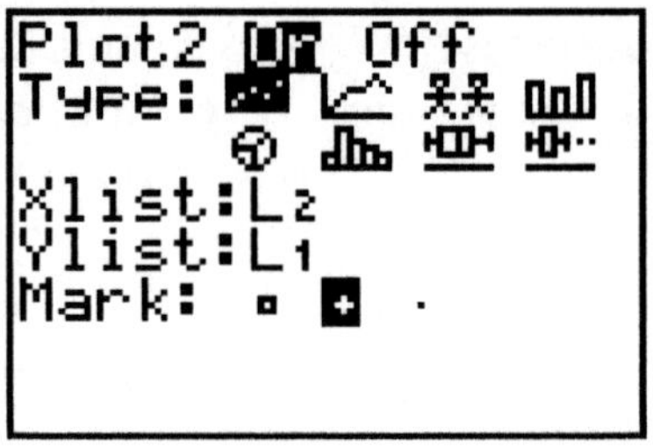

Analyzing the data: Finding a linear regression

1. Press [2nd] [STAT] [◀] to move the cursor to the **CALC** menu.

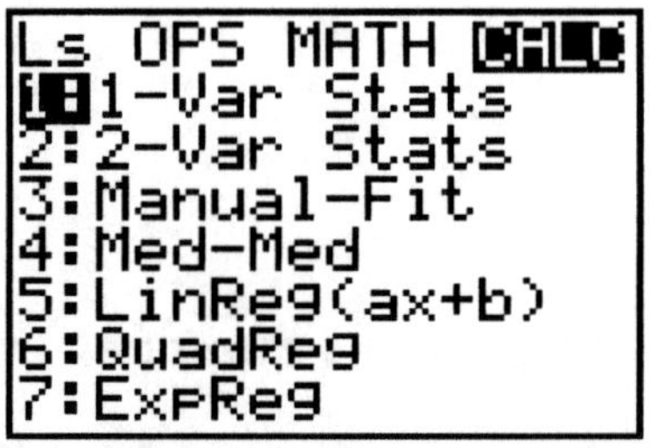

2. Select **5:LinReg(ax+b)** by pressing **5**.

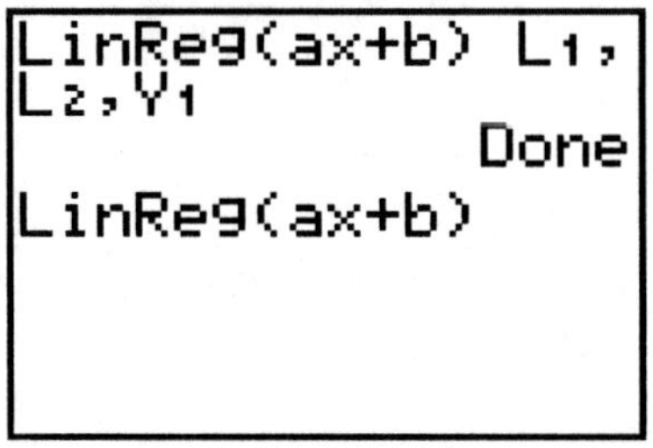

3. Press [2nd] [STAT] **2:L2** [,] [2nd] [STAT] **1:L1** [,].

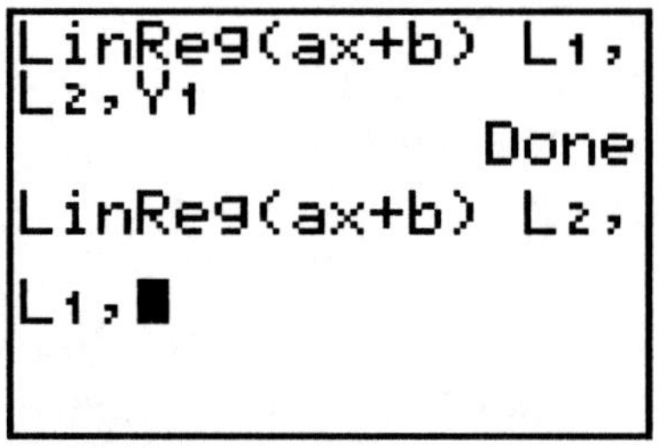

4. Press [2nd] [VARS].

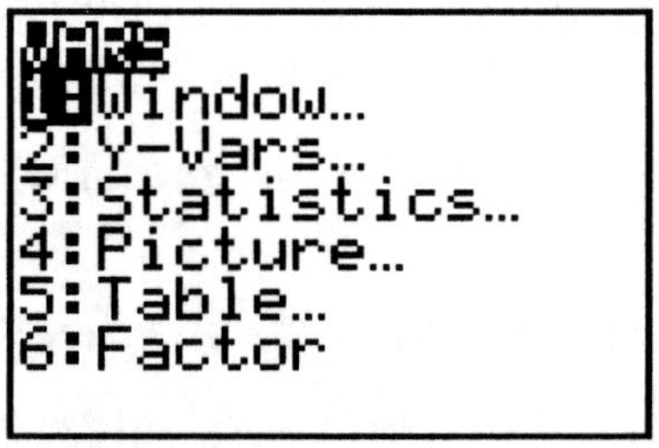

5. Select **2:Y-Vars** by pressing **2**.

6. Select **2:Y2** by pressing **2**.

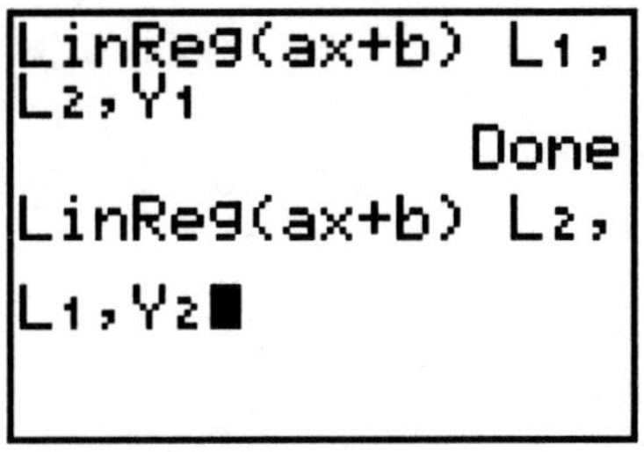

7. Press [ENTER] to calculate the linear regression and paste the function in **Y2**.

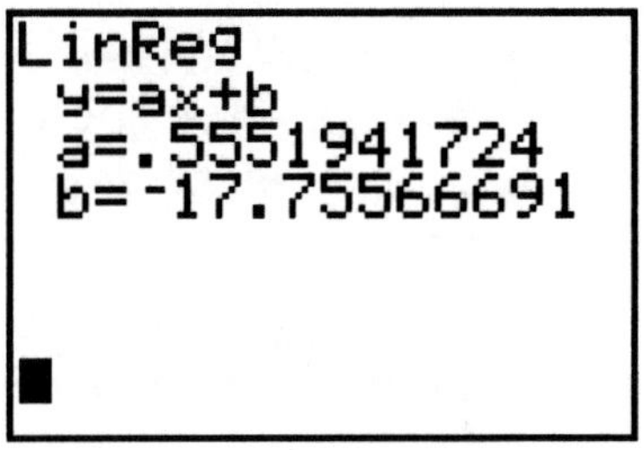

8. Press [GRAPH] to see the linear regressions.

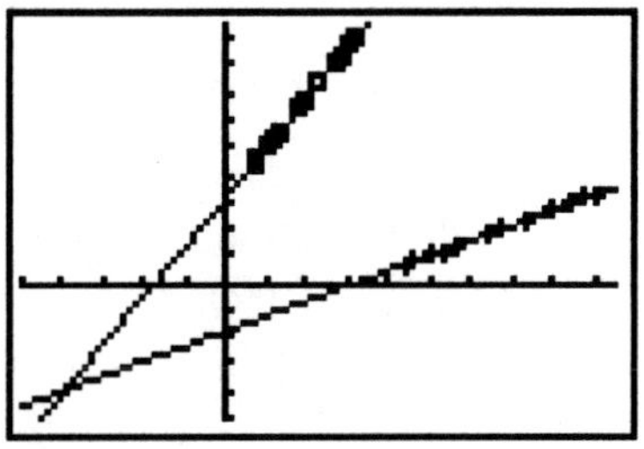

Answer Part II question 1 on the **Data Collection and Analysis** page.

Observe the following statements about the point of intersection of equations **Y1** and **Y2**:

- The *x* and *y* value of the point of intersection must satisfy *both* equations.
- Since one function (**Y2**) was obtained by switching the *x* and *y* values of the other function (**Y1**), *x* must equal *y* at the point of intersection.

Identify the point of intersection.

9. Press [2nd] [TBLSET]. Type [(-)] **45**. Press [▼] **1** to set the **ΔTbl** value.

   The table will allow you to examine increasing *x* values, starting at -45 degrees and the corresponding *y* values for both linear models that you plotted.

10. Press [2nd] [TABLE]. If necessary, use [▲] and [▼] to scroll the table.

    Examine the table carefully. Note that when rounding off to the nearsest whole number, there is one value that is the same for **X, Y1** and **Y2**. For this example, the coordinates of that point are (-40,-40).

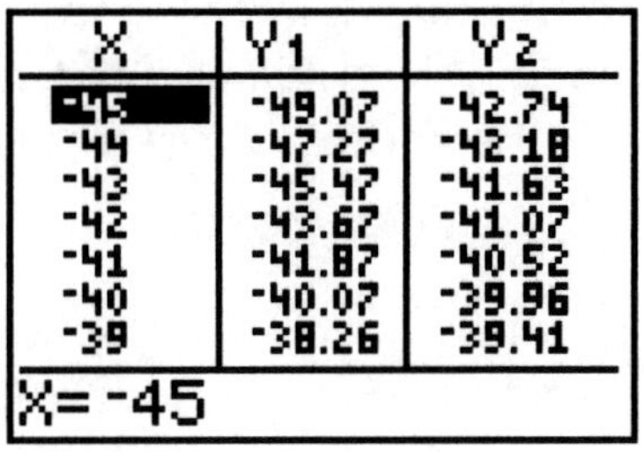

To verify the coordinates graphically, you will plot a vertical line (*x*=-40) and a horizontal line (*y*=-40).

11. Press [DRAW]. Select **3:Horizontal** by pressing **3**. Type the *x* value from Step 10, then press [ENTER]. (For this example, the value is -40.)

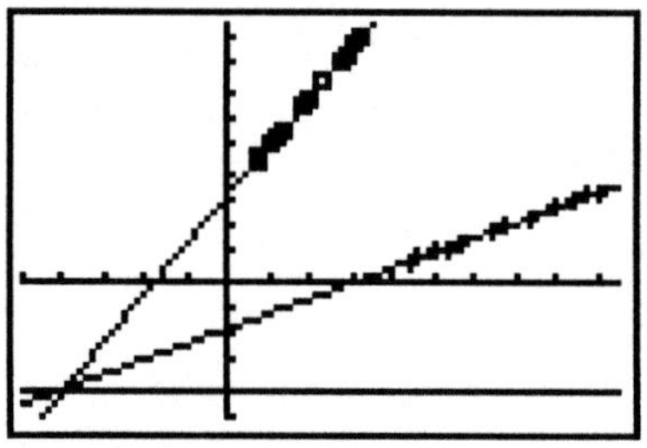

12. Press [2nd] [QUIT] [DRAW]. Select **4:Vertical** by pressing **4**. Type the **Y1** or **Y2** value from Step 10, then press [ENTER]. (For this example, the value is -40.)

    Note: *In both linear models, the horizontal line (y=-40), and the vertical line (x=-40) all meet at the same point (-40, -40).*

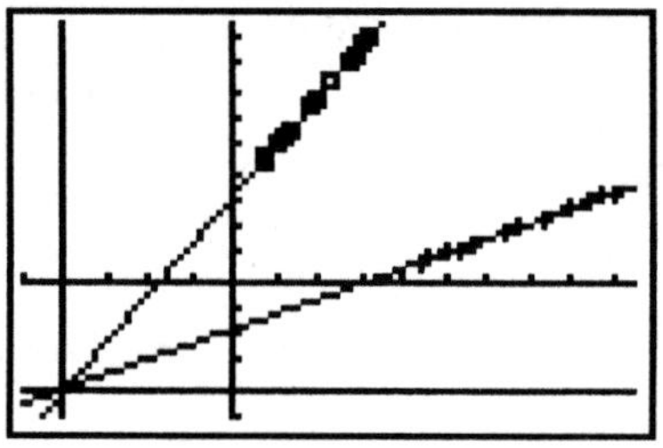

Add a graph of the function, y=x, to the graph.

13. Press [Y=]. Press [▼] to move the cursor to **Y3**. Press [x] so that the equation is **Y3 = X**.

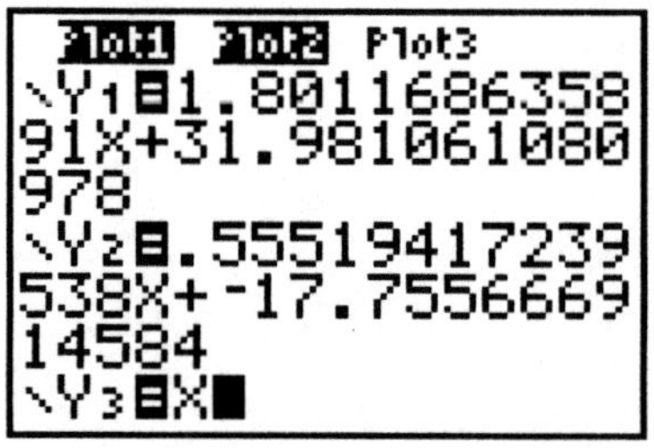

14. Press [GRAPH] to see the plot.

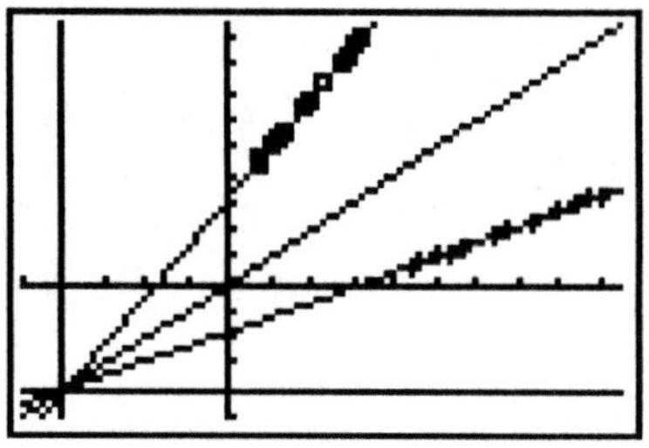

Answer Part II questions 2 - 5 on the **Data Collection and Analysis** page.

Collecting the data — Part III

You have learned about the relationship between the two units of temperature measurement. You will now explore how graphical analysis of climate data can be used to compare different ecosystems. You will compare rainfall and temperature in three locations, the tundra (Fairbanks, Alaska), the desert (Las Vegas, Nevada), and the tropical rain forest (San Jose, Costa Rica). Climate data for the three ecosystems is provided.

Setting up the TI-73

Before starting your data collection, make sure that the TI-73 has the STAT PLOTS turned OFF, Y= functions turned OFF or cleared, the MODE and FORMAT set to their defaults, and the lists cleared. See the Appendix for a detailed description of the general setup steps.

Entering the data in the TI-73

1. Press [LIST].

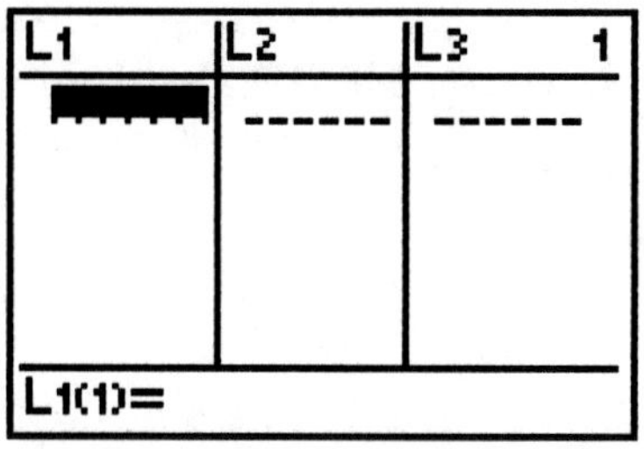

2. Enter the climate data in lists **L1** through **L6** according to the chart that follows.

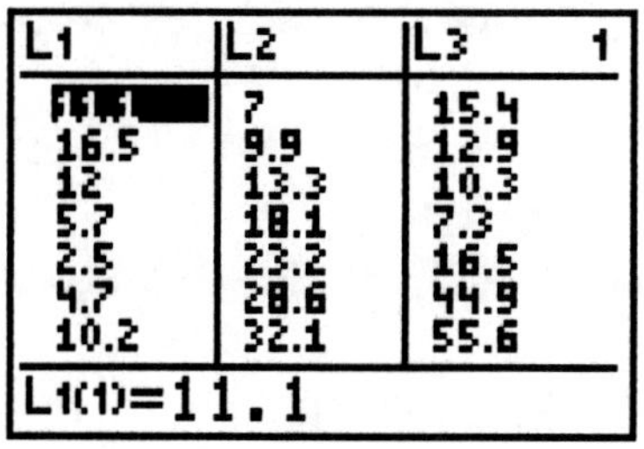

	Desert		Tundra		Tropical Rain Forest	
Month	**Rainfall (mm)**	**Temperature (°C)**	**Rainfall (mm)**	**Temperature (°C)**	**Rainfall (mm)**	**Temperature (°C)**
	L1	**L2**	**L3**	**L4**	**L5**	**L6**
Jan	11.1	7.0	15.4	-21.1	6.9	27.2
Feb	16.5	9.9	12.9	-18.1	2.7	27.9
Mar	12.0	13.3	10.3	-10.4	6.1	28.5
Apr	5.7	18.1	7.3	-1.0	32.8	28.7
May	2.5	23.2	16.5	8.5	199.2	27.9
Jun	4.7	28.6	44.9	14.6	240.0	27.2
Jul	10.2	32.1	55.6	16.1	183.0	26.9
Aug	13.3	30.9	52.9	13.2	243.1	26.9
Sep	10.6	26.6	32.3	7.0	308.7	26.5
Oct	6.3	19.6	21.8	-4.0	253.0	26.6
Nov	6.2	11.9	18.7	-15.3	118.7	26.6
Dec	14.2	7.5	21.1	-19.7	32.7	26.8

Source: Reprinted with permission from WorldClimate (www.worldclimate.com).

3. Move the cursor to the top of **L6** to highlight it. Press [▸] to move to the top of the seventh, unnamed, data list.

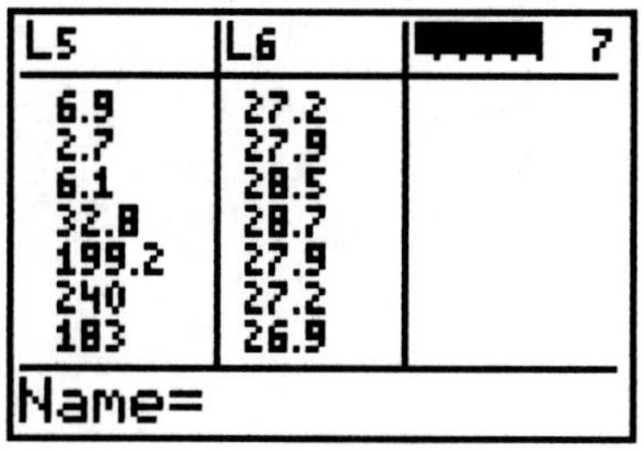

L5
L6
7
6.9
2.7
6.1
32.8
199.2
240
183
27.2
27.9
28.5
28.7
27.9
27.2
26.9
Name=

4. Name the list **MONTH** by pressing [2nd] [TEXT], moving the cursor to each letter of the name **MONTH**, and pressing [ENTER].

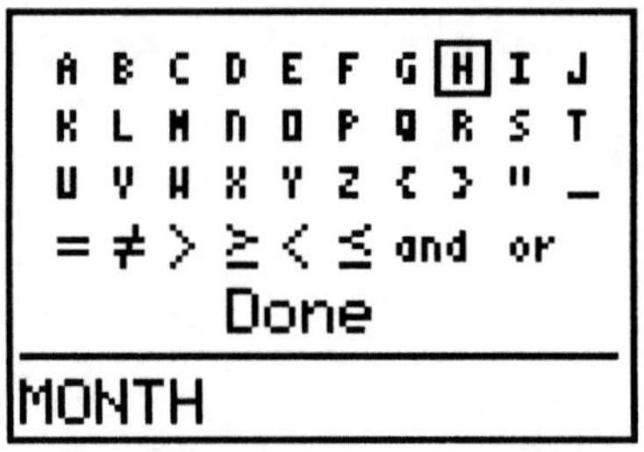

5. Use the cursor keys to move the cursor to **Done.** Press [ENTER] to exit the Text menu.

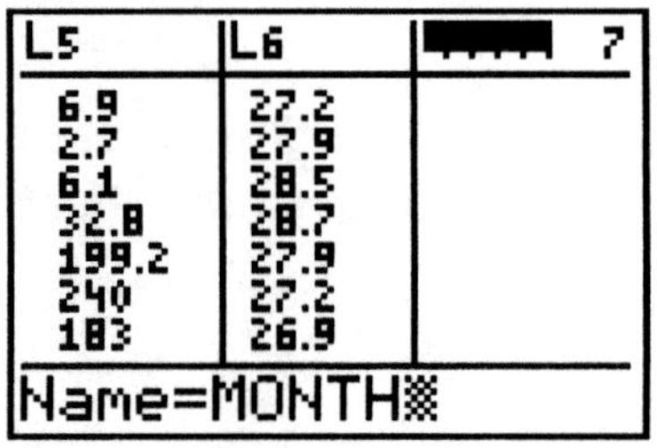

6. Press [ENTER] to save the list name.

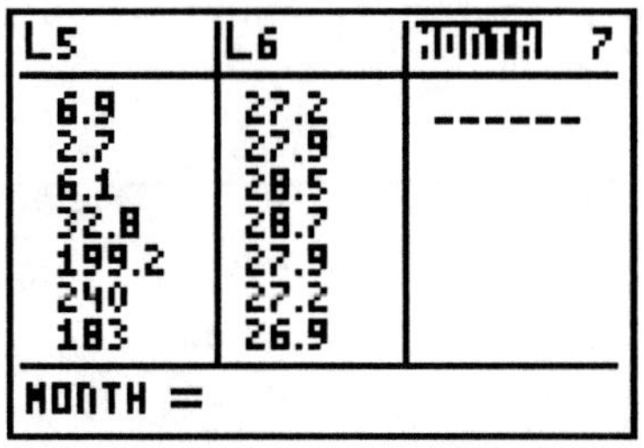

7. Press [▼] to move the cursor down in the data cell and enter the numbers **1** through **12** for each of the 12 months.

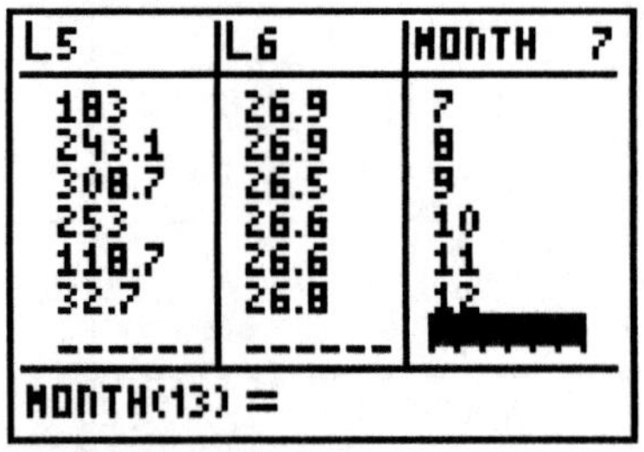

8. Press [2nd] [PLOT]. Select **1:Plot1** by pressing **1** or [ENTER].

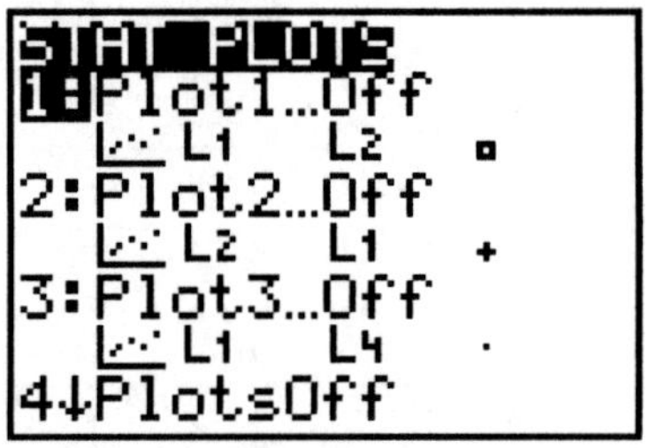

9. Set up the plot as shown by pressing [ENTER] [▼] [▶] [ENTER] [▼] [2nd] [STAT] **7:MONTH** [▼] [2nd] [STAT] **2:L2** [▼] [ENTER].

__Note__: The list, __MONTH__, may not be in position 7 on the TI-73. Use [▲] and [▼] to move the cursor to the desired list, and press [ENTER] to select that list.

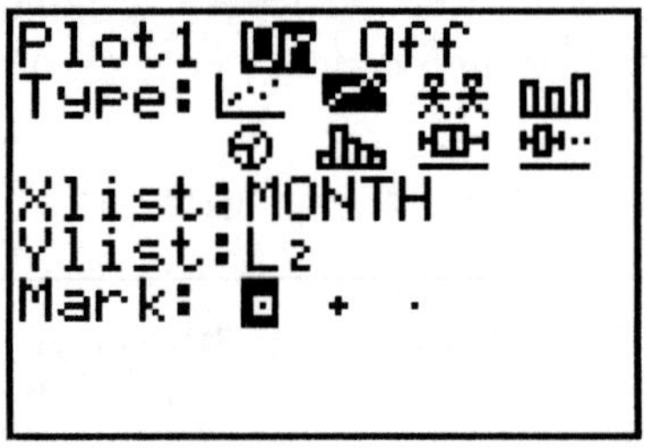

10. Press [ZOOM]. Select **7:ZoomStat** by pressing 7 to see a graph of monthly average temperatures for the desert.

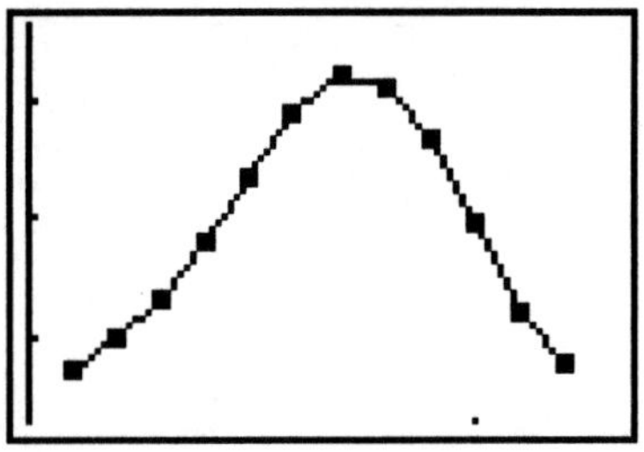

Answer Part III questions 1 and 2 on the **Data Collection and Analysis** page.

Graphing the data: Setting up a box-and-whisker plot

The type of analysis that you did with the desert temperature data could also be done with the other data from the table. Another way to analyze the data is by examining a box-and-whisker plot.

1. Press [2nd] [PLOT]. Select **1:Plot1** by pressing **1** or [ENTER].

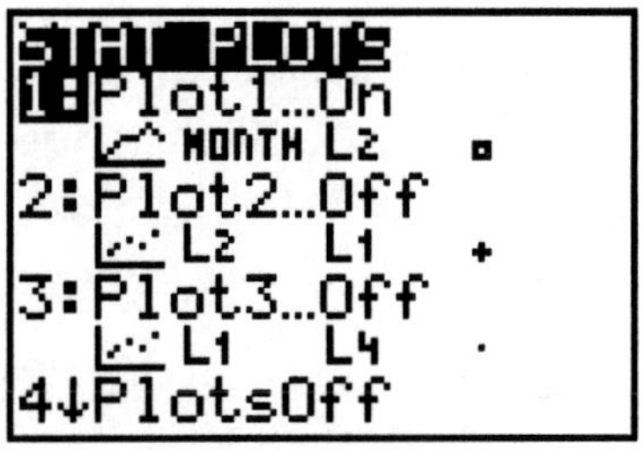

2. Set up the plot as shown by pressing [ENTER] [▼] [▶] [▶] [▶] [▶] [▶] [▶] [ENTER] [▼] [2nd] [STAT] **2:L2** [▼] **1**.

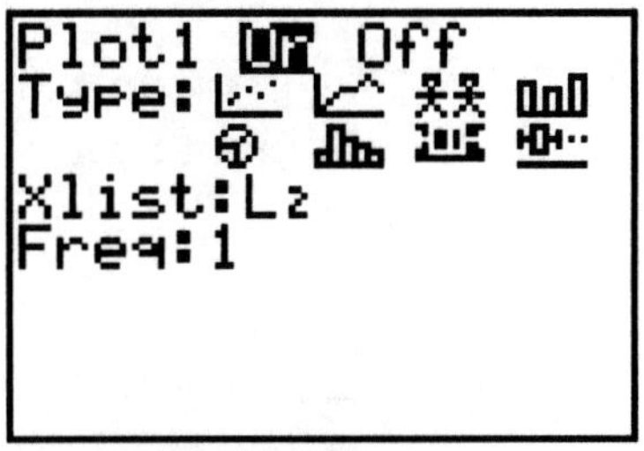

3. Press [2nd] [PLOT]. Select **2:Plot2** by pressing **2**.

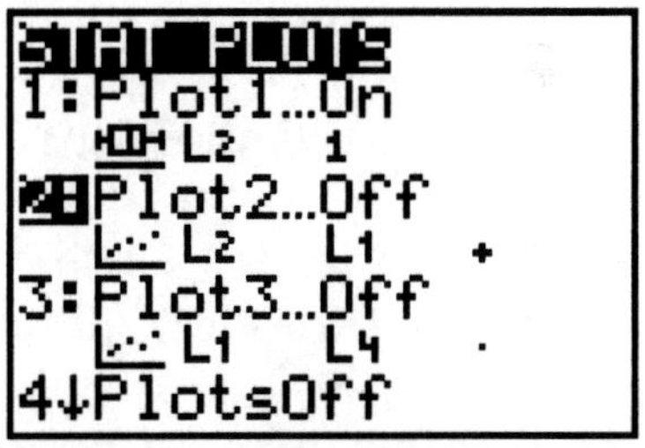

4. Set up the plot as shown by pressing [ENTER] [▼] [▶] [▶] [▶] [▶] [▶] [▶] [ENTER] [▼] [2nd] [STAT] **4:L4** [▼] **1**.

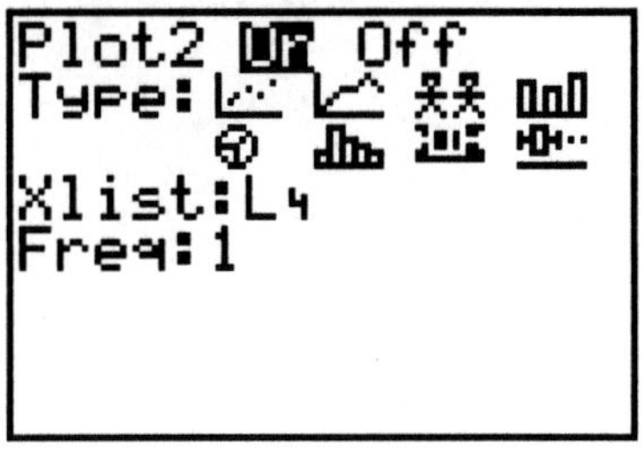

5. Press [2nd] [PLOT]. Select **3:Plot3** by pressing **3**.

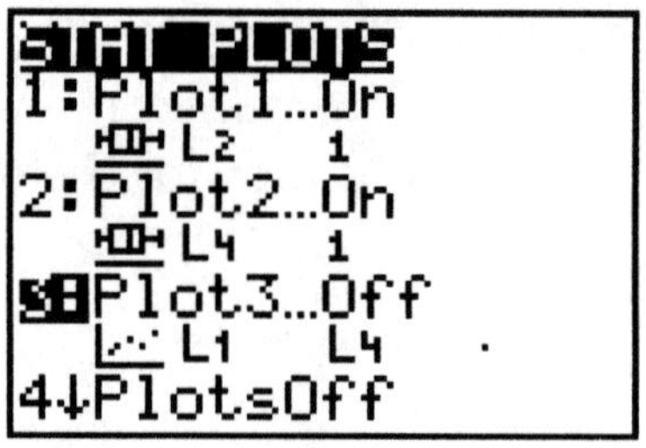

6. Set up the plot as shown by pressing [ENTER] [▼] [▶] [▶] [▶] [▶] [▶] [▶] [ENTER] [▼] [2nd] [STAT] **6:L6** [▼] 1.

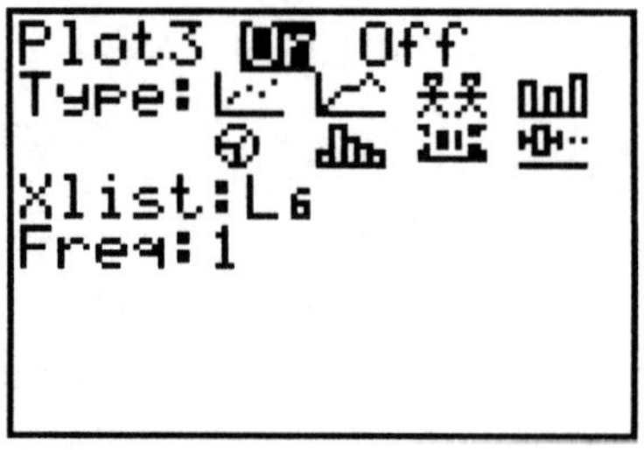

7. Press [WINDOW] and set the window values as shown. **Do Not Change the ΔX Value**.

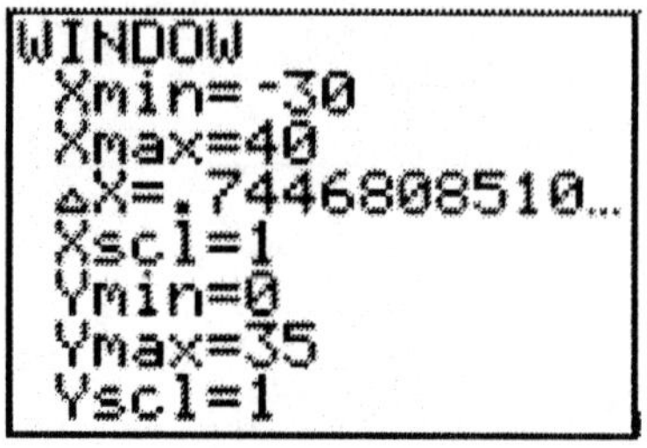

8. Press [GRAPH] to compare the three box-and-whisker plots.

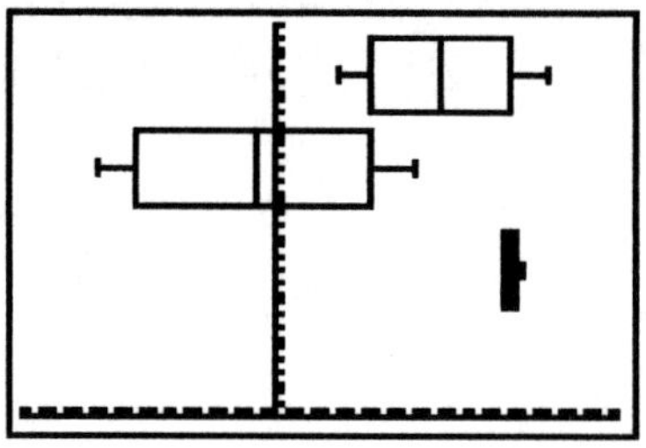

Answer Part III questions 3 through 5 on the **Data Collection and Analysis** page.

9. Repeat Steps 1-8 using the rainfall data for the desert (**L1**) for **Plot1**, the Tundra (**L3**) for **Plot2**, and the Tropical Rain Forest (**L5**) for **Plot3**.

Answer Part III questions 6 and 7 on the **Data Collection and Analysis** page.

Extension

A useful tool to investigate ecosystems is a climagraph. A climagraph is a plot of Temperature (on the *y*-axis) versus Rainfall (on the *x*-axis). Each point on the plot represents the average rainfall and temperature for a given month.

Construct climagraphs for the desert (**L1**, **L2**), the tundra (**L3**, **L4**), and the rain forest (**L5**, **L6**). Since climagraphs are cyclical, repeat (re-enter) the first value of each list as the 13th element of the same list. (The first and 13th values of each list will be identical.)

What information does a climagraph give you that you cannot obtain from looking at the temperature or rainfall data in isolation?

Data Collection and Analysis

Name ______________________

Date ______________________

Activity 13: Do You Have a Temperature?

Collecting the data

Temperature (°C)												
Temperature (°F)												

Analyzing the data — Part I

1. What is the equation for the linear regression in **Y1**? (You can see the equation by pressing [Y=].)

____

2. What does *y* represent in this equation? What does *x* represent in this equation?

____

3. How might this equation be used to convert the temperature in one scale to the temperature in another scale?

____

____

4. How does this conversion equation compare to that in a math or science textbook?

____

____

5. What is the *slope* of this function?

____

6. What does the *slope* tell you? (Use the words degrees Fahrenheit and degrees Celsius to answer this question.)

____

____

Analyzing the data — Part II

1. Examine the general appearance of the two functions (**Y1** and **Y2**) and answer the following questions:

   a. In which quadrant do the **Y1** and **Y2** functions intersect? ____________

   b. Describe the symmetry of the two functions.

   ____

   ____

   c. What function serves as a line of symmetry for the two functions? (Provide an equation with your answer.)

   ____

   ____

2. Rounding off to the nearest whole number, what is the point of intersection of functions **Y1** and **Y2**? ______________________________

3. How do the *x* and *y* values compare at the point of intersection?

   ____

   ____

4. Press [Y=] to see the equations in **Y1** and **Y2**. Show that the point of intersection is correct for these two equations by substituting the *x* and *y* values. Show your work.

   ____

   ____

   ____

5. Based on the appearance of the three functions, confirm your answers to questions 1b and 1c. Explain.

   ____

   ____

   ____

   ____

Analyzing the data — Part III

1. Describe the temperature pattern for the desert over a one-year period.

2. How would the graph of the average monthly temperature in the desert change from year-to-year? Explain.

3. Which ecosystem is most stable according to temperature? How does the box-and-whisker plot show this?

4. Which ecosystem has the lowest median temperature value during the course of the year?

5. Which ecosystem has the highest median temperature value during the course of the year?

6. Which ecosystem has the highest median rainfall amount during the course of the year?

7. Which ecosystem has the lowest median rainfall amount during the course of the year?

Teacher Notes

Activity 13

Do You Have a Temperature?

Objectives

- To graphically represent and analyze climate data
- To use linear regressions to understand the relationship between temperatures as measured in the Fahrenheit and Celsius scale
- To use linear regressions to understand conversion factors
- To use technology to find a linear regression

Materials

- TI-73 graphing device
- CBL 2™ data collection device (optional)
- Climate data for different ecosystems
- Cold cup with ice water
- Hot cup with boiling water
- Rubber band
- Watch with a second hand
- Celsius thermometer and Fahrenheit thermometer
- Two temperature probes (per CBL 2) (optional)

Preparation

- This activity shows that conversions between scientific units of measurement are linear. Similar conversions could be done with centimeters to inches and pounds to kilograms. Since the *y*-intercept for the above conversions is 0, the slope is the conversion factor. In the examples used in this activity, the conversion equations are:

  $F = 1.8\ C + 32$ *and* $C = 0.56\ F - 17.8$

  Graphs of these equations form mirror images on both sides of the $Y = X$ line. They intersect at the one temperature where degrees C = degrees F (-40 degrees).

- If you do not want to collect temperature data, use the monthly average temperature provided for Las Vegas, Nevada, or look up similar data for your location.

Answers to Data Collection and Analysis

Collecting the data

Sample data — Las Vegas, Nevada:

Month	J	F	M	A	M	J	J	A	S	O	N	D
Temperature (°C)	7.0	9.9	13.3	18.1	23.2	28.6	32.1	30.9	26.6	19.6	11.9	7.5
Temperature (°F)	44.6	49.8	55.9	64.6	73.8	83.5	89.8	87.6	79.9	67.3	53.4	45.5

Source: Reprinted with permission from WorldClimate (www.worldclimate.com).

- If the students collect their own data using thermometers, do not expect the equation to fit as perfectly as using the Las Vegas, Nevada, data in the table. There are several reasons for this:
  - The thermometer bulbs are not in the same place.
  - Slight errors in reading the thermometers may occur.
  - Convection currents within the container result in slight temperature differences.
  - Students may not take the reading at the same time.

  There are benefits to discussing the discrepancies that occur when using real world data.
- The web page mentioned above is an excellent source of data for cities all over the world. You may want to make this a web-based activity.

Analyzing the data — Part I

1. What is the equation for the linear regression in **Y1**? (You can see the equation by pressing [Y=].)

   For the Las Vegas data, $Y = 1.801X + 31.981$.

2. What does y represent in this equation? What does x represent in this equation?

   The y represents the temperature in degrees Fahrenheit. X represents the temperature in degrees Celsius.

3. How might this equation be used to convert the temperature in one scale to the temperature in another scale?

   If you know the temperature in degrees Celsius, substitute the value for x and simplify to get the temperature in degrees Fahrenheit.

4. How does this conversion equation compare to that in a math or science textbook?

   F = 1.8 C + 32 (Textbook Equation).

5. What is the *slope* of this function?

   1.8

6. What does the *slope* tell you? (Use the words degrees Fahrenheit and degrees Celsius to answer this question.)

   The slope tells you that for each increase in a Celsius degree, the Fahrenheit temperature rises 1.8 degrees.

Analyzing the data — Part II

1. Examine the general appearance of the two functions (**Y1** and **Y2**) and answer the following questions:

   a. In which quadrant do the **Y1** and **Y2** functions intersect?

   The **Y1** *and* **Y2** *functions intersect in the third quadrant.*

   b. Describe the symmetry of the two functions.

   The two functions are inverses of each other.

   c. What function serves as a line of symmetry for the two functions? (Provide an equation with your answer.)

   They are symmetrical with respect to the line, Y = X.

2. Rounding off to the nearest whole number, what is the point of intersection of functions **Y1** and **Y2**?

   (-40, -40)

3. How do the *x* and *y* values compare at the point of intersection?

   The x and y values are the same at the point of intersection.

4. Press [Y=] to see the equations in **Y1** and **Y2**. Show that the point of intersection is correct for these two equations by substituting the *x* and *y* values. Show your work.

   By substituting - 40 for the x and y values of the two equations, and simplifying, the equality will be shown.

5. Based on the appearance of the three functions, confirm your answers to questions 1b and 1c. Explain.

   The equations all intersect at (-40, -40).

Analyzing the data — Part III

1. Describe the temperature pattern for the desert over a one-year period.

   The temperature pattern starts low, rises, and then falls to the same level.

2. How would the graph of the average monthly temperature in the desert change from year-to-year? Explain.

   The graphs would be approximately the same although slight variations may occur. This is because average temperatures do not vary a great deal.

3. Which ecosystem is most stable according to temperature? How does the box-and-whisker plot show this?

   The Tropical Rain Forest is the most stable according to temperature. The plot is the most narrow.

4. Which ecosystem has the lowest median temperature value during the course of the year?

   The Tundra has the lowest median temperature value during the course of the year.

7. Which ecosystem has the highest median temperature value during the course of the year?

   The Tropical Rain Forest has the highest median temperature value during the course of the year.

8. Which ecosystem has the highest median rainfall amount during the course of the year?

   The Tropical Rain Forest has the highest median rainfall amount during the course of the year.

9. Which ecosystem has the lowest median rainfall amount during the course of the year? (This is hard to see from the graph.)

   The Desert has the lowest median rainfall amount during the course of the year, but the Tundra is a close second.

EXPLORATIONS

Activity 14

The Closer I Get to You

Objectives

- To find a rational function
- To find the *x* value of a function, given the *y* value
- To find the *y* value of a function, given the *x* value
- To use technology to plot data

Materials

- TI-73 graphing device
- Small mirror (one per group)
- Adding machine paper (at least 500 cm per group)
- Masking tape and markers
- Meter stick (one per group)

Introduction

If you drop a ball on the ground, it will return to you along the same path in which it was released. If you throw the ball so that it hits the floor at an angle, the ball will return in the opposite direction, keeping the same angle with the floor. Light follows a similar pattern. When light rays strike a surface, the ray that strikes the surface is called the *incident ray*, and the ray that is reflected is called the *reflected ray*. The law that describes this phenomenon is called the *Law of Reflection*. This law states that the angle of incidence is equal to the angle of reflection. (See diagram below.)

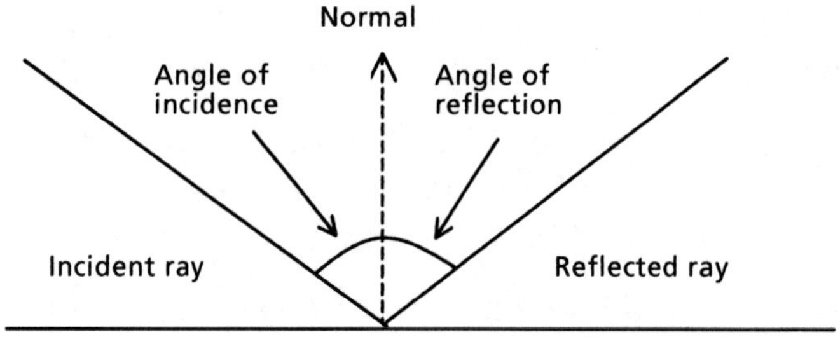

To explore the *Law of Reflection*, hold a mirror in the palm of your hand parallel to the ground and at a level that allows you to look into the mirror (see diagram on the next page). Find an object or classmate that is above your eye level (for example, a poster on the wall, a word on the blackboard, or a person in the room). Walk away from the object and observe what happens as you look in the mirror. Walk towards the object and observe what happens as you look in the mirror. Your discovery might lead you to an interesting observation that can be used to find the height of an object.

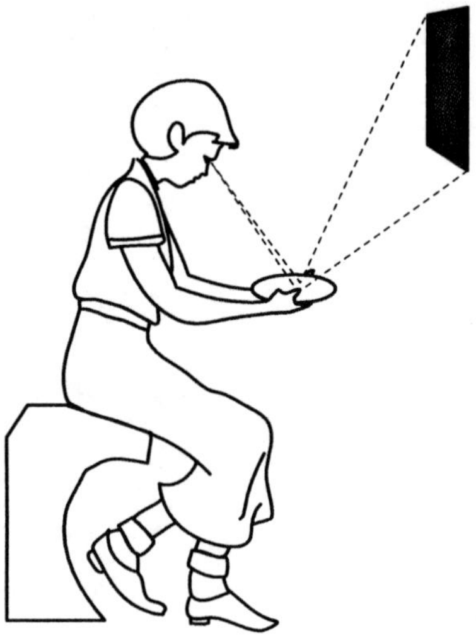

Problem

How can you use the *Law of Reflection* to find the height of objects that are impractical to measure using conventional techniques? Can you use a mirror to find the height of a classmate?

Collecting the data

1. Obtain two pieces of adding machine paper from your teacher. One should be slightly longer than one meter in length, and the second should be about four meters in length. Mark 0 (zero) at one end of each strip of paper. (Write the number at least 3 centimeters tall.) Measure, mark and label each strip every 10 centimeters, starting at zero, until you reach the end of each strip of paper.
2. Tape the 1-meter paper strip to the wall, making sure 0 is positioned on the floor.
3. Tape the 4-meter paper strip on the floor, with 0 in the same position as the paper strip on the wall. Make sure this strip is perpendicular to the wall.
4. Obtain a mirror from the teacher. Using a washable marker, draw a 2 cm x 2 cm square in the center of the mirror. Place the mirror on the paper on the floor at the 20 cm mark. The square should be centered on the 20 cm mark.
5. Select one of the group members to be the spotter, one as the marker, and one as the scribe. Measure the *eye-level* height in centimeters of the spotter.
6. Instruct the spotter to face the paper strip on the wall. Ask the spotter to move away from the wall until the reflection of the numeral 10 can be seen in the mirror. Mark the position of the spotter's toe on the paper strip on the floor. Record this position in the table on the **Data Collection and Analysis** page. (See diagram that follows.)

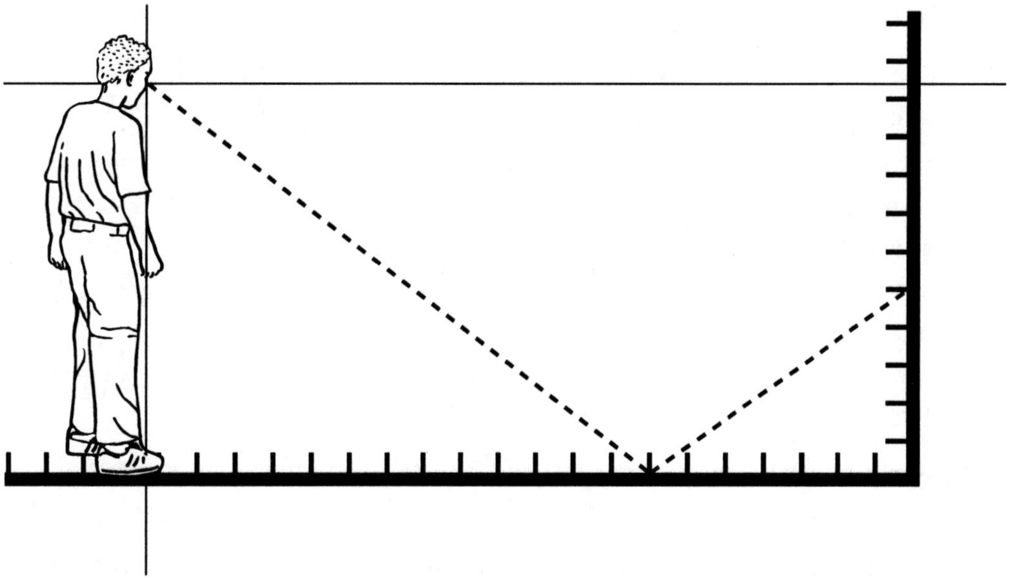

7. Ask the spotter to walk slowly towards the paper strip on the wall until the reflection of the numeral 20 can be seen in the mirror. Mark the position of the spotter's toe on the paper strip on the floor. Record this position in the table on the **Data Collection and Analysis** page.

   Note: *The spotter's distance from the wall is the only variable that changes.*

8. Continue this procedure until the spotter has seen the reflection of all the numerals on the paper strip.

Setting up the TI-73

Before starting your data collection, make sure that the TI-73 has the STAT PLOTS turned OFF, Y= functions turned OFF or cleared, the MODE and FORMAT set to their defaults, and the lists cleared. See the Appendix for a detailed description of the general setup steps.

Entering the data in the TI-73

1. Press [LIST].

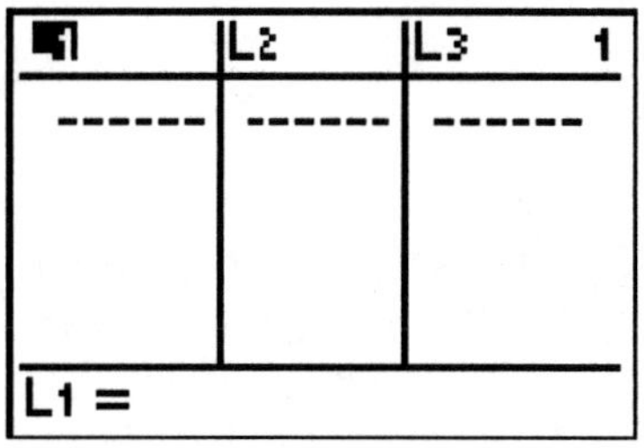

2. Enter the data for the height on the wall in **L1**.

3. Enter the data for the distance from the wall in **L2**.

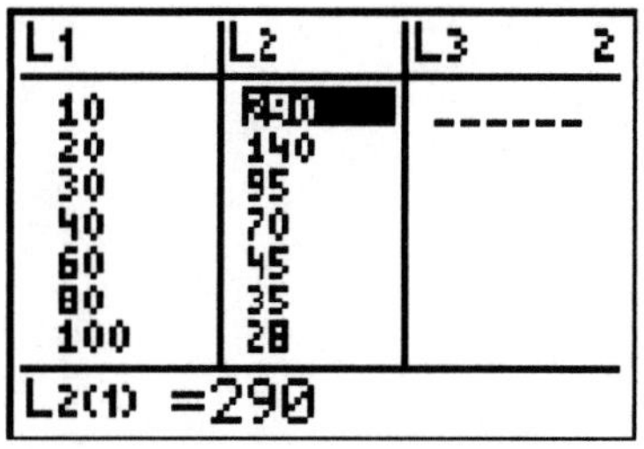

Setting up the window

1. Press [WINDOW] to set up the proper scale for the axes.

2. Set the **Xmin** value by identifying the minimum value in **L1**. Choose a number that is less than the minimum.

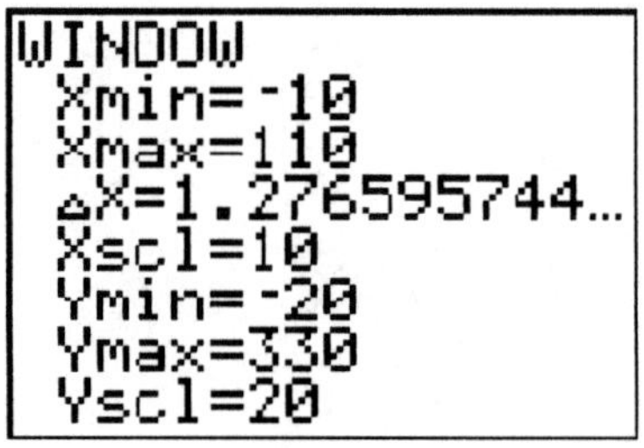

3. Set the **Xmax** value by identifying the maximum value in each list. Choose a number that is greater than the maximum. **Do Not Change the ΔX Value.** Set the **Xscl** to **10**.

4. Set the **Ymin** value by identifying the minimum value in **L2**. Choose a number that is less than the minimum.

5. Set the **Ymax** value by identifying the maximum value in **L2**. Choose a number that is greater than the maximum. Set the **Yscl** to **20**.

Graphing the data: Setting up a scatter plot

Plot a scatter plot using the data in the table on the **Data Collection and Analysis** page.

1. Press [2nd] [PLOT]. Select **1:Plot1** by pressing **1** or [ENTER].

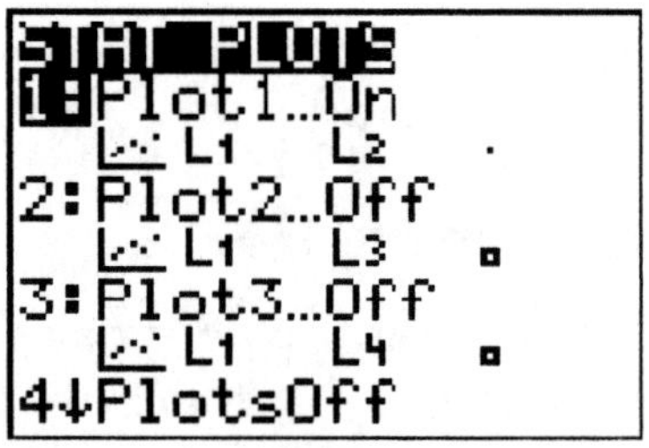

2. Set up the plot as shown by pressing [ENTER] [▼] [ENTER] [▼] [2nd] [STAT] **1:L1** [▼] [2nd] [STAT] **2:L2** [▼] [ENTER].

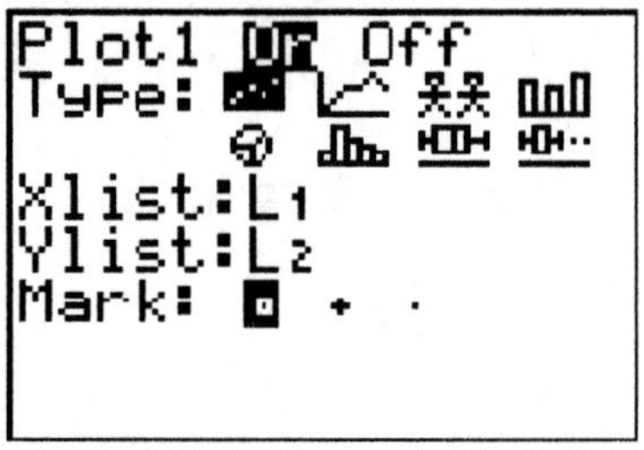

3. Press [TRACE] to see the plot. Press [▶] to view the data points.

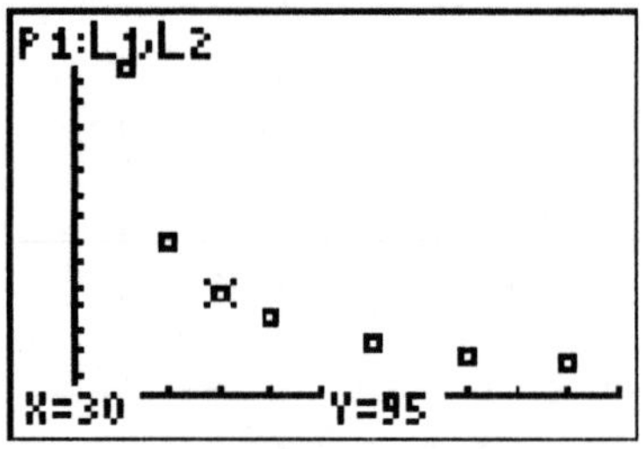

Analyzing the data: Finding a function

The function that models this data is a rational function. Examine the information below to determine a rational model for the data.

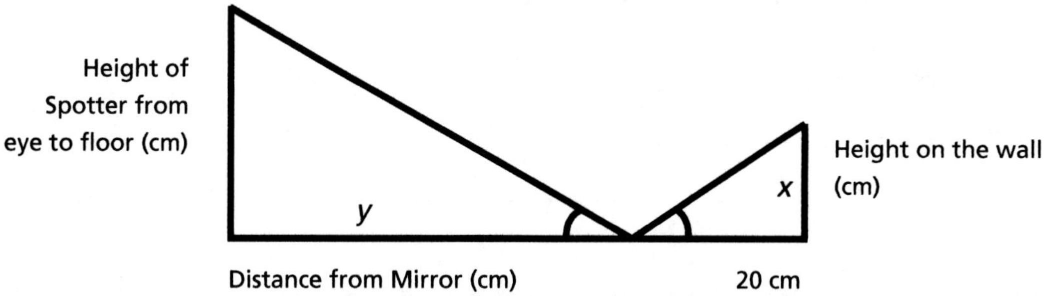

Rational equation:

$$\frac{y}{\text{Height to Spotter (cm)}} = \frac{20\text{ (cm)}}{x}$$

or

$$y = \frac{(\text{Height of Spotter}) \times (20)}{x}$$

Using the spotter's height, enter the rational model in *y*.

1. Press [Y=] [(] **1 4 6** [)] **2 0** [÷] [x].

   ***Note**: In this example, the spotter's height was 146 cm. Replace this value with the actual height of your spotter.*

   Record the equation on the **Data Collection and Analysis** page.

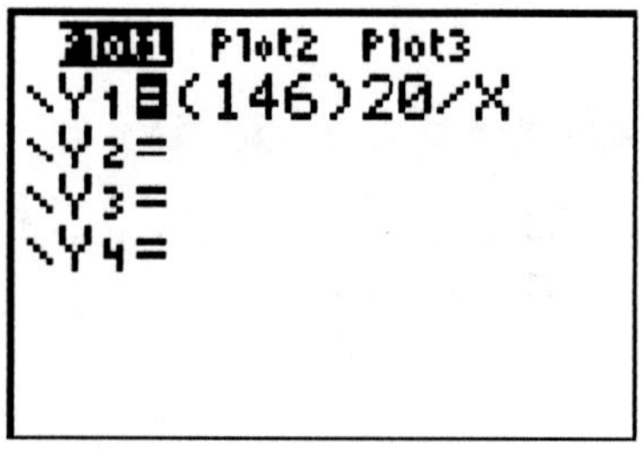

2. Press [GRAPH] to see the graph of the function.

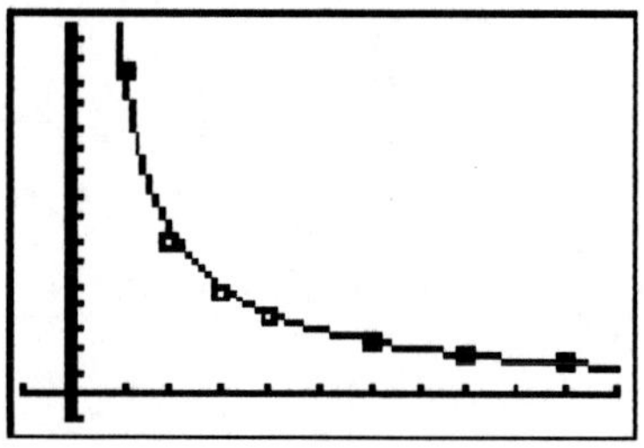

Answer questions 1 through 6 on the **Data Collection and Analysis** page.

Data Collection and Analysis

Name ______________________

Date ______________________

Activity 14: The Closer I Get to You

Collecting the data

Record your data in the table below.

Height (cm)	Distance from mirror (cm)
10	
20	
30	
40	
60	
80	
100	

Equation: ______________________________

Analyzing the data

1. Use your equation to find the distance from the mirror if the height visible in the mirror is:

   a. 45 cm ______________________

   b. 110 cm ______________________

2. Use your equation to find the height on the wall if the spotter stands the following distances from the mirror:

   a. 100 cm ______________________

   b. 40 cm ______________________

   c. 200 cm ______________________

3. What will happen to the graph of the data if the spotter is taller?

   ____

   ____

4. What will happen to the graph of the data if the spotter is shorter?

____

____

5. Determine the height of a classmate. Place a piece of masking tape on the wall to mark your classmate's height. Walk away from the mirror until you see the masking tape in the mirror. Use the equation to find your classmate's height.

____

6. Find a classmate's height, in centimeters, using the meter stick. Determine where you would have to stand to see the top of the classmate's head.

____

Teacher Notes

Activity 14

The Closer I Get to You

Objectives

- To find a rational function
- To find the *x* value of a function, given the *y* value
- To find the *y* value of a function, given the *x* value
- To use technology to plot data

Materials

- TI-73 graphing device
- Small mirror (one per group)
- Adding machine paper (at least 500 cm per group)
- Masking tape and markers
- Meter stick (one per group)

Preparation

- Use a marker for the numbering on the tape.
- Make sure students are not bending to see the numeral in the mirror.

Answers to Data Collection and Analysis questions

Collecting the data

Sample data:

Height (cm)	Distance from mirror (cm)
10	290
20	135
30	90
40	62
60	41
80	28
100	19

Find an equation for the data by using the formula.

For the sample data, $y = \dfrac{146(20)}{x}$.

Analyzing the data

1. Enter your equation in **Y1** in the TI-73. Use your equation to find the distance from the mirror if the height visible in the mirror is:

a. 45 cm. *Answers will vary. For the sample data, stand approximately 65 cm from the mirror.*

b. 110 cm. *Answers will vary. For the sample data, stand approximately 27 cm from the mirror.*

2. Use your equation to find the height on the wall if the spotter stands the following distances from the mirror:

a. 100 cm. *Answers will vary. For the sample data, the height on the wall is 29.2 cm.*

b. 40 cm. *Answers will vary. For the sample data, the height on the wall is 73 cm.*

c. 200 cm. *Answers will vary. For the sample data, the height on the wall is 14.6 cm.*

3. What will happen to the graph of the data if the spotter is taller?

The graph will appear to be shifted up if the spotter is taller.

4. What will happen to the graph of the data if the spotter is shorter?

The graph will appear to be shifted down if the spotter is shorter.

5. Determine the height of a classmate. Place a piece of masking tape on the wall to mark your classmate's height. Walk away from the mirror until you see the tape in the mirror. Use the equation to find your classmate's height.

Answers will vary.

6. Find a classmate's height, in centimeters, using the meter stick. Determine where you would have to stand to see the top of the classmate's head.

Answers will vary.

Appendix

Before starting data collection using the TI-73, check the following settings.

Checking the MODE settings

The MODE settings control how the TI-73 displays and interprets numbers and graphs. The MODE should be set to the default setting of the TI-73. The term *default setting is* calculator terminology that describes a screen where all of the items or functions selected are in the leftmost column.

If the MODE setting is not set to the defaults:

1. Press [MODE] to display the mode settings.

2. Press [▼] until the cursor is on the line that does not have the leftmost option highlighted.

3. Press [ENTER] to change the setting.

4. Continue this process until you have selected all of the options that are in the leftmost column.

The mode is now set to the defaults.

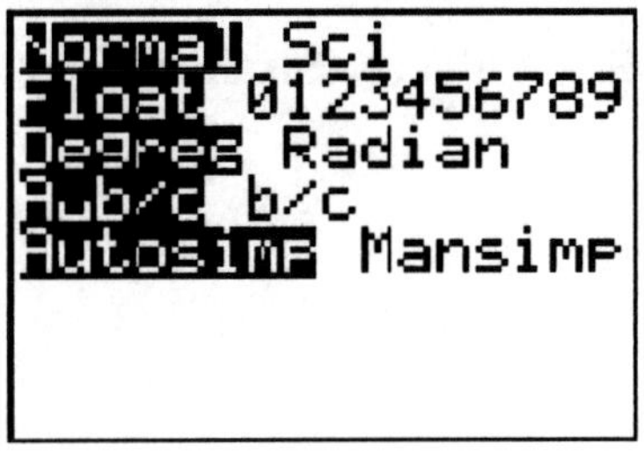

Checking the Y= Editor

The TI-73 can store up to 4 functions to the variables **Y1** through **Y4**. Make sure that there are no functions stored in the Y= Editor.

1. Press [Y=] to display the Y= Editor.

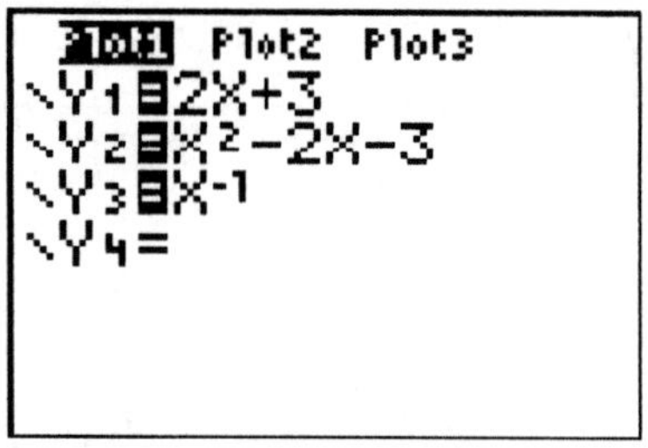

2. If any functions are stored in the Y= Editor, press [▼] to move the cursor to the line that contains a function. Press [CLEAR] to remove functions.

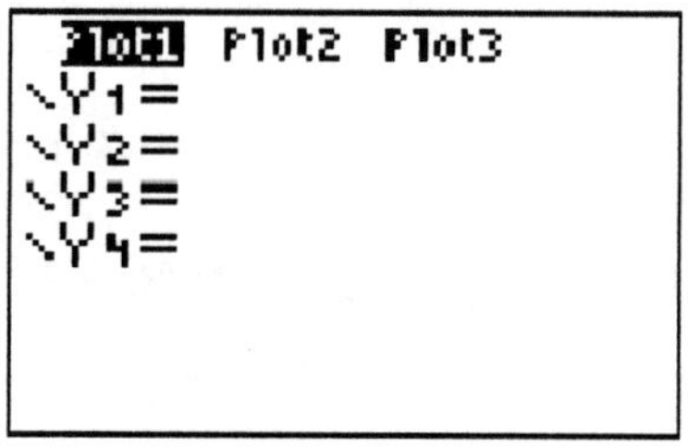

Checking the WINDOW Settings

The WINDOW sets the viewing rectangle for the function's graph.

1. Press [WINDOW] to display the window.

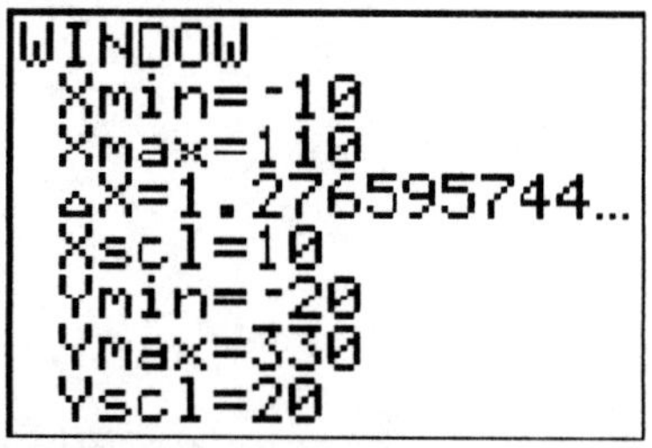

2. Press [ZOOM].

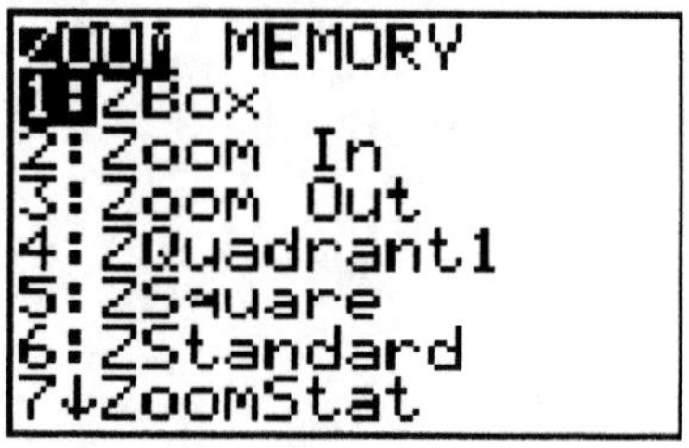

3. Select **6:ZStandard** by pressing **6**.

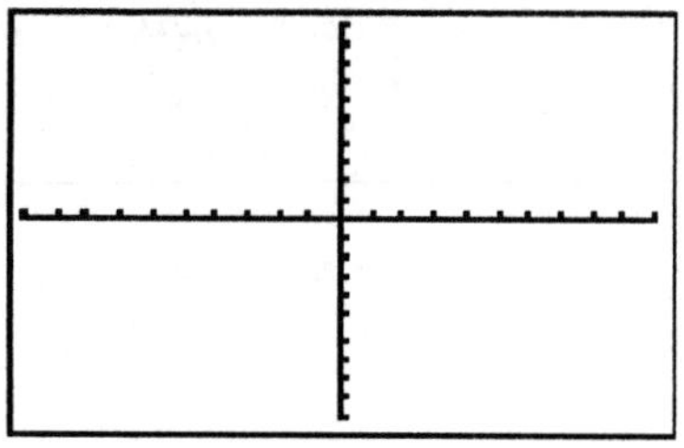

4. Press [WINDOW] to view the default settings.

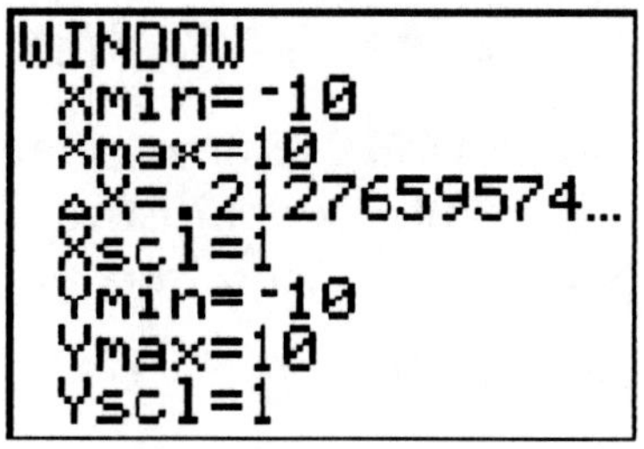

Checking the FORMAT settings

Make sure the **FORMAT** menu is set to the defaults. (Refer to the **Checking the MODE Settings** section for an explanation of defaults.)

1. Press [2nd] [FORMAT] to display the **FORMAT** window.
2. Press [▼] until the cursor is on the line that does not have the default (leftmost) option highlighted.
3. Press [ENTER] to change the setting.
4. Continue this process until you have selected all of the options that are in the leftmost column.

   The format is now set to the defaults.

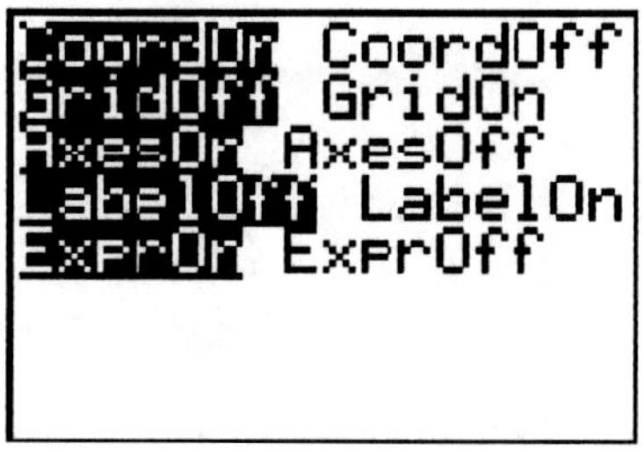

Checking the STAT PLOTS settings

The **STAT PLOTS** menu controls the way in which data that has been entered in the calculator's statistical lists are plotted and/or displayed. There are several ways to turn all of the plots **OFF**. This section discusses two of those methods.

Method I

1. Press [Y=].
2. Check to see if any of the plots are selected.

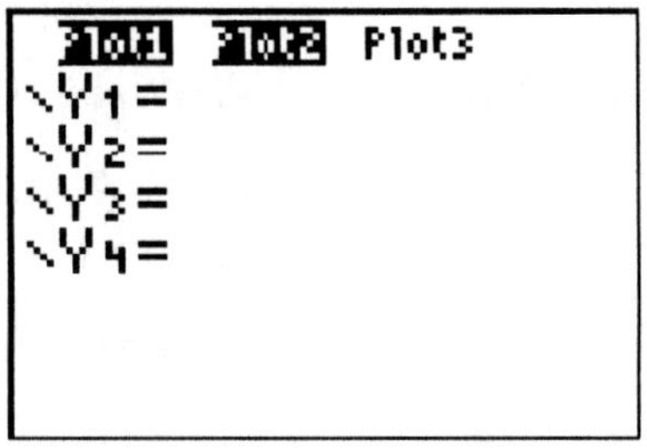

3. If a plot is highlighted, press [▲] and [▶], as necessary, to move the cursor to the highlighted plot.
4. Press [ENTER] to turn the plot **OFF**.

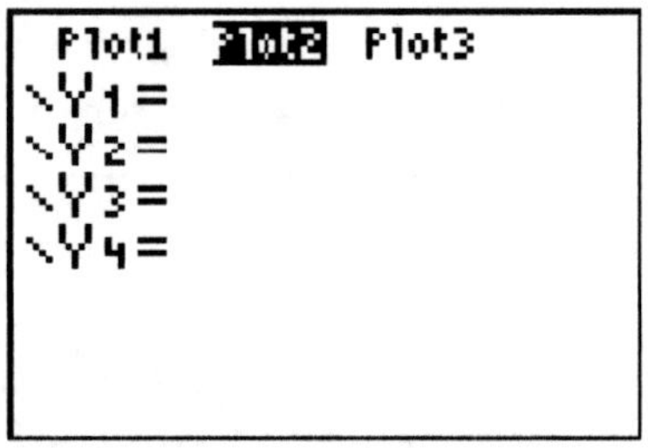

5. Repeat Steps 3 and 4 until all plots are off.

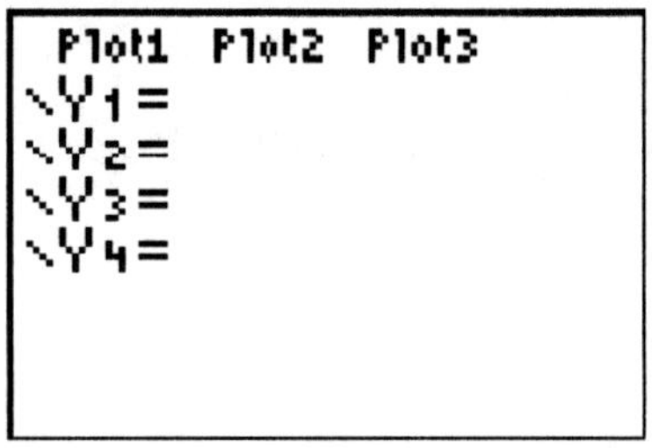

Method II

1. Press [2nd] [PLOT] to display the **STAT PLOTS** menu.

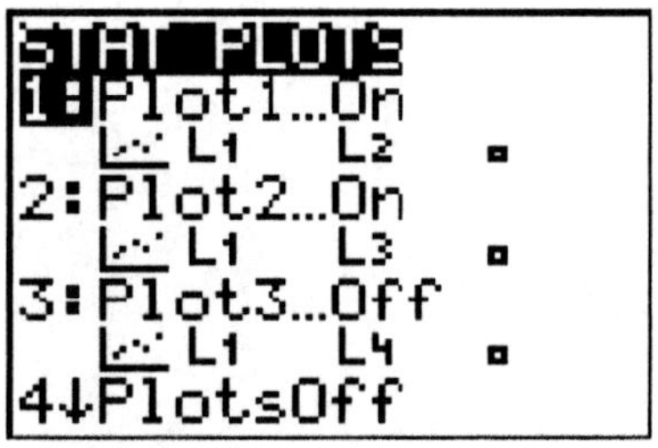

2. If any of the plots are **ON**, select **4:PlotsOff** by pressing **4**.

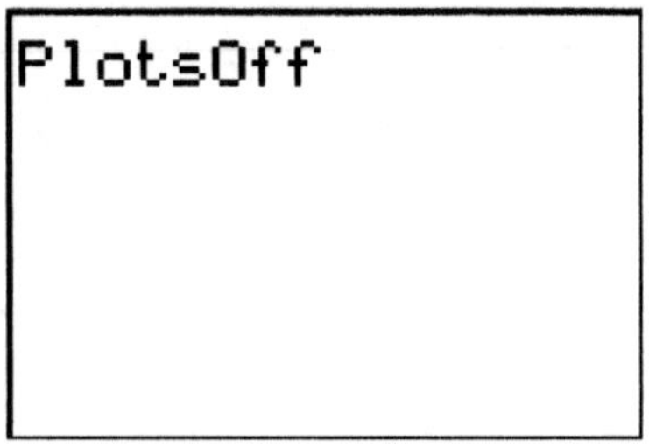

3. Press [ENTER] to turn the plots off. The TI-73 screen displays the word **Done**.

4. Press [2nd] [PLOT] to display the **STAT PLOTS** menu. All the plots should read **OFF**.

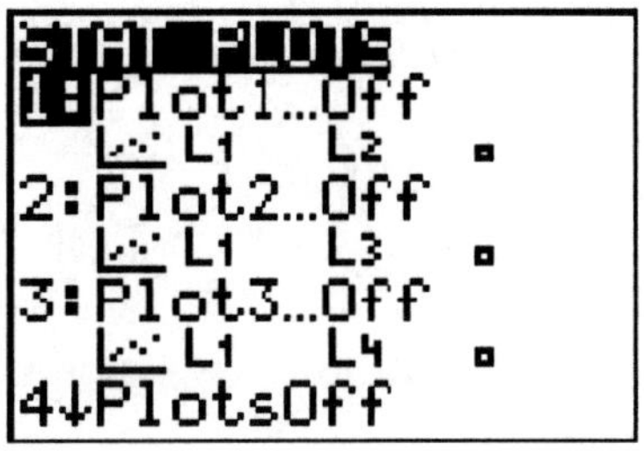

Using reset defaults to reset the MODE, WINDOW, FORMAT, and STAT PLOTS settings

The **MODE, WINDOW, FORMAT,** and **STAT PLOTS** can be reset simultaneously to default mode. Resetting the defaults on the TI-73 turns the plots **OFF**, sets the **MODE** and **FORMAT** to default settings, and sets a standard **WINDOW**. However, resetting the defaults *will* turn off but not clear out the functions in the Y= Editor.

Use the **Reset** menu to reset the **MODE, WINDOW, FORMAT,** and **STAT PLOTS**:

1. Press [2nd] [MEM].

2. Select **7:Reset** by pressing **7**.

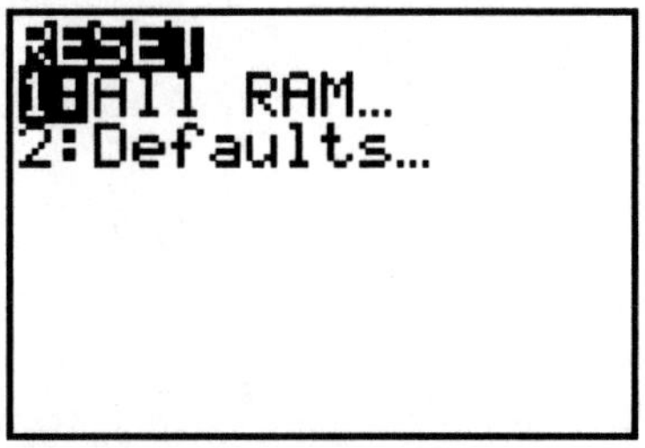

3. Select **2:Defaults** by pressing **2**.

4. Select **2:Reset** by pressing **2**.

5. Press [CLEAR] to clear the Home screen.

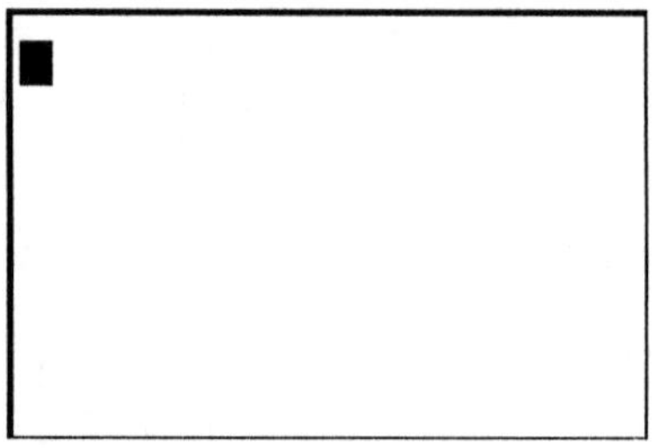

6. If you check the Y= Editor (press [Y=]), you will notice that the functions have been turned **OFF**. (The equal sign is not highlighted.)

   If you want to turn the functions **ON**, move the cursor over the equal sign and press [ENTER].

   For data collection, leave the functions OFF. If you paste a regression equation in a function, **Y**, which already contains an equation, the regression equation will overwrite the existing equation.

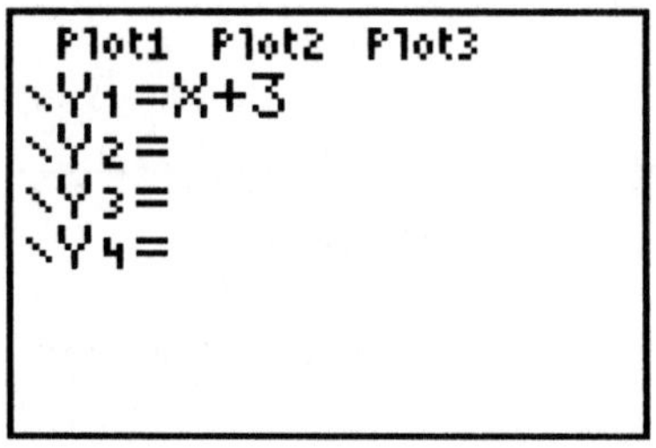

Clearing lists

Lists can be cleared in several different ways on the TI-73. This section discusses three ways in which this can be done.

Method I

1. Press [2nd][LIST].

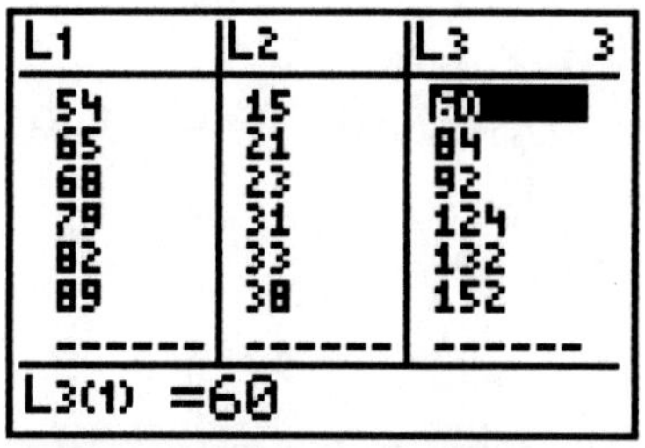

2. Use the cursor keys (◄ ► ▼ ▲) to move the cursor to the top of the list you want to clear.

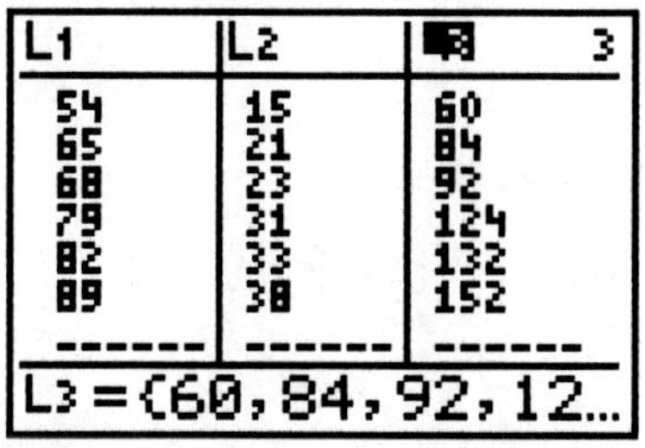

3. Press [CLEAR] [ENTER] to clear the selected list.

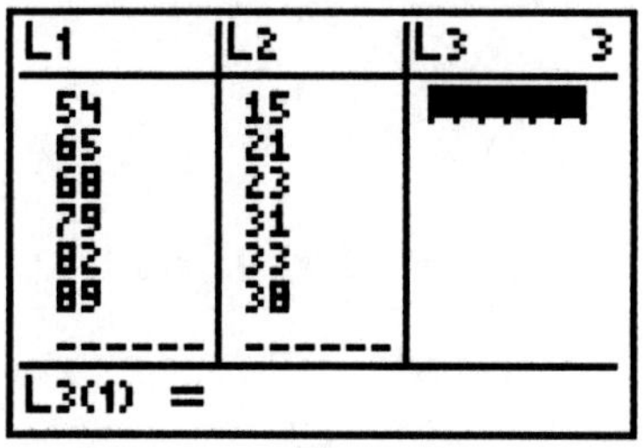

Method II

1. Press [2nd] [STAT] ► to move the cursor to the **OPS** menu.

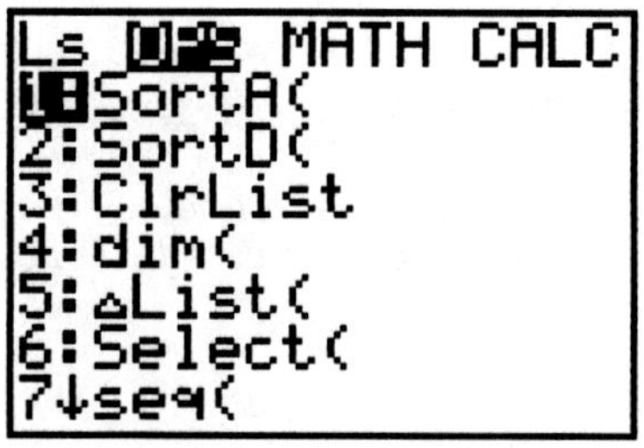

2. Select **3:ClrList** by pressing **3.**

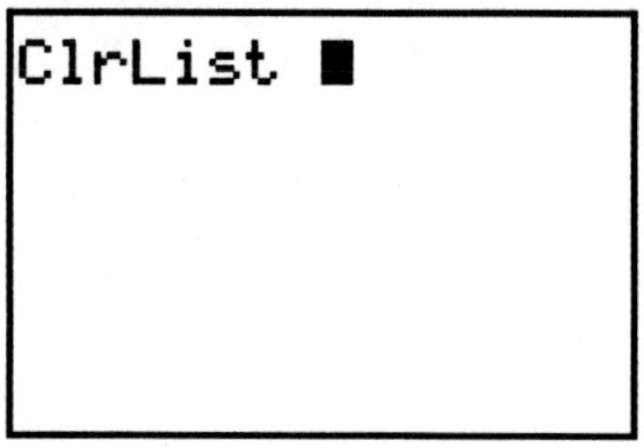

3. Press [2nd] [STAT] **1:L1** [,] **2:L2**.

   Note: *This will clear* ***L1*** *and* ***L2.*** *Additional lists can be included. Lists need to be separated by commas.*

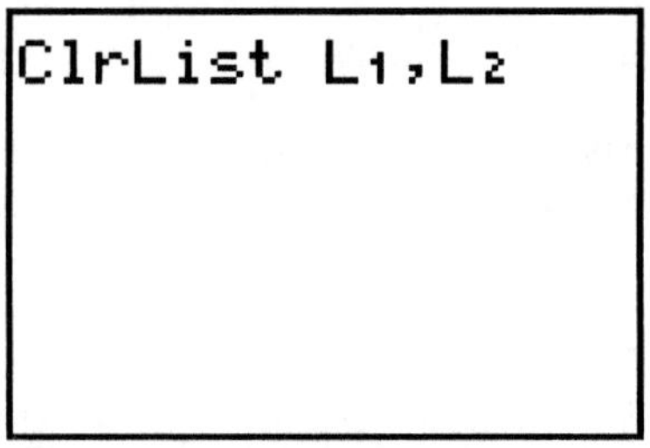

4. Press [ENTER] to clear the lists.

Method III

1. Press [2nd] [MEM].

2. Select 6:**ClrAllLists** by pressing **6.**

3. Press [ENTER] to clear ALL lists.

   Note: *This method clears all lists, even those that do not appear on the list editor, but it does not clear the list names.*

EXPLORATIONS

Index

E

F

G

H

I

L

M